Basics in

FINFISH AND SHELLFISH PHYSIOLOGY

AUTHORS

Ms. A Arunashri obtained her degree of Bachelor of Fisheries Science (B.F.Sc) in the year 2019 from Dr. M.G.R. Fisheries College and Research Institute, Ponneri, under TNJFU Tamil Nadu. She cleared AIEEA PG entrance examination in the year 2019. She completed her Postgraduate degree of Master of Fisheries Science (M. F. Sc) in Fish Physiology and Biochemistry in the year 2021 form ICAR- Central Institute of Fisheries Education. She also cleared AIEEA PhD exam in the year 2021. She cleared ASRB NET in Fish Nutrition Discipline in the year 2023. She has also written several articles and book chapters. She had nine years of experience in fisheries field particularly six years of experience in Fish Nutrition, Physiology and Biochemistry Discipline.

Mr. Sai Kishore Potluri obtained his Bachelor of Fisheries Science (B.F.Sc.) degree in 2020 from the College of Fishery Science, Muthukur, under SVVU, Tirupati, Andhra Pradesh. He cleared the ICAR-AIEEA PG entrance examination in 2020 and went on to complete his Master of Fisheries Science (M.F.Sc.) in Fish Nutrition and Feed Technology in 2022 from the ICAR-Central Institute of Fisheries Education, Mumbai. He also cleared ICAR-AIEEA PhD exam in 2022. He has cleared ASRB NET in Fish Nutrition discipline in the year 2023. With eight years of experience in the fisheries field, including four years specializing in Fish Nutrition, Physiology, and Biochemistry, he has contributed significantly to the fisheries science community through several articles and book chapters.

Ms. P Arya, MFSc. took her Bachelor of Fisheries Science(B.F.Sc) degree in the year 2016 from Kerala University of Fisheries and Ocean Studies, Panangad, Kochi, Kerala. She cleared AIEEA PG entrance examination in the year 2016 and joined ICAR- Central Institute of Fisheries Education, Mumbai for her Post graduation. She completed her Master of Fisheries Science in Fish Physiology and Biochemistry during the year 2018. Later on she has joined as a PhD Scholar in Fish Physiology and Biochemistry department at ICAR-CIFE Mumbai. She cleared ASRB NET in Fish Nutrition Discipline in 2018. Currently she is appointed as Assistant Professor at Kerala University of Fisheries and Ocean Studies, Panangad, Kochi, Kerala.

Basics in
FINFISH AND SHELLFISH PHYSIOLOGY

– Author –

A Arunashri

– Co-Authors –

Sai Kishore Potluri

P Arya

ASTRAL

Publishing Authors Across the Globe

2025

Daya Publishing House®

A Division of

Astral International Pvt. Ltd.

New Delhi – 110 002

ISBN: 978-93-5919-934-4

Published by : **Daya Publishing House®**
A Division of
Astral International Pvt. Ltd.
– ISO 9001:2015 Certified Company –
4736/23, Ansari Road, Darya Ganj
New Delhi-110 002
Ph. 011-43549197, 23278134
e-mail: info@astralint.com
Website: www.astralint.com

Printed at : **Replika Press Pvt. Ltd.**

PREFACE

This book, "Basics in Finfish and Shellfish Physiology," is specifically written with Bachelor of Fisheries Science undergrads in mind. Although there are numerous works on advanced finfish physiology accessible, this book aims to provide UG students with a foundational understanding of the physiology of finfish and shellfish. This book is based on the ICAR syllabus, which is followed by almost all the fisheries colleges across India. To provide general understanding, a brief introduction to Fish Nutrition and Biochemistry is also included in the book. This book will be helpful for those who are preparing for JRF and Fisheries competitive examinations. The Authors have briefly covered the fundamentals of Finfish and Shellfish physiology for students at the beginner level.

CONTENTS

UNIT II

UNIT III

UNIT IV

UNIT V

UNIT I

CHAPTER 1

WATER AS A BIOLOGICAL MEDIUM

1.1. What is water?

Water is essential for every living being and without water, life would not exist. The molecular formula of water is **H_2O**. Covalent bonds join the two hydrogen atoms and one oxygen atom to form the water molecule. Water exists in three forms depending on the surrounding temperature. At ambient temperature, it will be in liquid for; at 0°C it will be in the solid state; and at 100°C it will be in the gaseous state. Water covers 71% of the earth's surface, where oceans hold 97% of surface water; glaciers, polar ice caps holds 2.4%, and other land surface water such as rivers, lakes, ponds holds 0.6%. Water is a major component of cells, typically forming between 70 and 95% of the mass of the cell. Humans are made of approximately 80% water by mass, and some soft-bodied creatures, such as jellyfish, are made of up to 96% water. **Fish** contains about **66-81%** of water.

1.2. Water cycle or hydrological cycle

12.1 Definition

The hydrologic cycle involves the **continuous circulation of water in the Earth-Atmosphere system.** At its core, the water cycle is the motion of the water from the ground to the atmosphere and back again.

1.2.2. Parts of the water cycle

Evaporation

Evaporation is the process of a liquid's surface changing to a **gas**. In the water cycle, liquid water (in the ocean, lakes, or rivers) evaporates and becomes water vapour. The water cycle's evaporation process is driven by the sun.

Condensation

Condensation is the process of a **gas changing to a liquid**. In the water cycle, water vapour in the atmosphere condenses and becomes liquid. Clouds form as water vapour condenses.

Precipitation

Precipitation describes any **liquid or solid water that falls** to earth due to condensation in the atmosphere. Precipitation includes rain, snow, and hail.

Transpiration

Transpiration is the process of **water** vapour being released **from plants** and soil.

Percolation

Percolation is the **downward movement of water** through soil layers due to gravity and capillary forces.

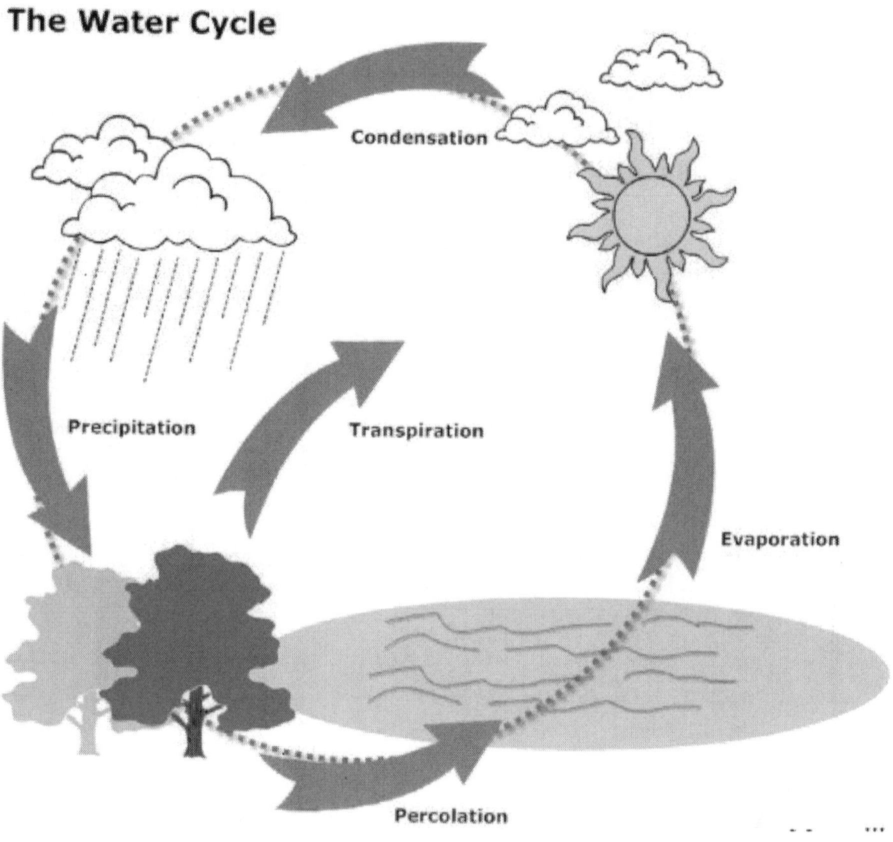

Fig 1.1: Water cycle

1.3. Water as a Habitat

Life originally evolved in an aqueous habitat, and although life now exists on land, it still provides a suitable habitat for a wide variety of life. Water is still an ideal medium for life, and there are many reasons behind this:

1.3.1. Water as a Major Cell Constituent

In most organisms, there is 60-90% water. The lowest water content can be found in plant seeds (20%) and the highest in jellyfish (99%). In cells, water is found mainly in the protoplasm, and here it plays vital roles in metabolism.

1.3.2. Water is the Universal Solvent

Convert food in soluble form (from starch to glucose) that can be transported and taken by various cells/tissues. It helps in the removal of metabolic waste products in the form of urine, sweat etc. It helps in respiration for the supply of oxygen from the heart to the body tissue and the intake of CO_2 from the body tissue the heart via blood. It helps in gas exchange, in aquatic habitat oxygen is dissolved in water, which is taken by the aquatic organisms

1.3.3. Water in the process of Transport

Blood transports nutrients, hormones, oxygen, waste products etc. within the animal body; similar to the sap transporting nutrients and other substances in plants. Both of these mediums for transport (blood and sap) are mainly water.

1.3.4. Water as a Reactant

The most basic example of this is photosynthesis. The equation for photosynthesis shows that water is a vital reactant in the reaction $6H_2O + 6CO_2 = C_6H_{12}O_6 + 6O_2$. Condensation, during the conversion of glucose to starch, water releases. Hydrolysis, during digestion, water requires the conversion of starch into glucose.

1.3.5. Water as a Mechanism of Support

Aquatic organisms have weaker skeletons than organisms living on land and the water's 'buoyancy effect' makes them 'lighter', thus giving them extra support.

1.3.6. Water as a Lubricant

Bones meet at joints, and lubrication must be provided to ensure the bones do not scrape against each other causing damage, enabling free, easy movement by reducing friction. A synovial membrane at joints encloses a synovial fluid, which acts as the lubricant. Water is the major constituent of this fluid and other bodily fluids such as:

- Cerebro-spinal fluid (Brain)
- Pleural fluid (Lungs)
- Aqueous and vitreous humours (eyes)
- Amniotic fluid (fetus)
- Mucus (gut)

1.3.7. Water in Sexual Reproduction

In most fish, external fertilization takes place in water. During fertilization, the sperm is often transported in a fluid medium known as semen, which contains mostly water.

1.3.8. Water as Temperature Controller

Water has a high specific heat capacity, with an approximate value of 4200J/kg°C, so the water temperature is not easily changed. A large mass of water, such as an ocean, will heat up slowly during the day and cool slowly at night, so its temperature does not change much. This provides an ideal habitat for marine organisms with only minor temperature variations. The high water content of cells gives them insulation and protects them from rapid temperature changes, thus helping to keep cells at a relatively constant optimum temperature.

Freshwater Ecosystem- Ponds, rivers, lakes, beels etc.

Marine Ecosystem- Ocean, sea etc.

1.4. Properties of water

A. Physical properties – Density, Buoyancy, Wave and current, Temperature, Salinity, Light and turbidity, Colour etc.

B. Chemical properties – DO, pH, Hardness, Alkalinity, CO_2 etc.

CHAPTER 2

GAS EXCHANGE

2.1. Introduction

2.1.1. Gas Exchange

Gaseous exchange is defined as the transfer of oxygen (O_2), carbon-di-oxide (CO_2), and, to a lesser extent, ammonia (NH_3) between the environment and the cell/tissue occurs at the site of use or production.

2.1.2. Respiration

Respiration consists of intake of oxygen (O_2) for metabolism/energy production and release of carbon dioxide (CO_2) into the environment as an end product of metabolism.

2.1.3. Types of respiration:

Respiration can be divided into two major types such as-

- External respiration
- Internal respiration

External respiration: The respiration in which gaseous exchange (O_2 & CO_2) takes place **between blood and water** (or air) through the medium of the respiratory organ is known as external respiration.

Internal respiration: The respiration in which the essential transfer of gasses **between blood and tissues or cells** of the body and brings about the release of energy is known as internal respiration.

The physiologically most important gases are oxygen, carbon dioxide and nitrogen. Their presence in the air is constant.

Component	%
Oxygen	20.95
Carbon dioxide	0.03

Component	%
Nitrogen	78.09
Argon	0.93

Animal	Respiratory organ
Jellyfish	No respiratory apparatus (simple **diffusion**)
Fulmonate snail	Diffusion lung
Sea cucumber	Water lungs
Fish	Tracheae (system of tubes that branch and lead to all parts of the body with no blood circulation)
Fish	Gill (formed due to evagination of the respiratory surface)

2.2. Respiratory Organs in Fishes

The **gills** are the primary organ by which gases are exchanged between the fish and the surrounding water. Through the gills, fish can absorb oxygen and give off carbon dioxide. Like the lungs, the gills have a large area for gaseous exchange. Some species have altered gills and other organs so that they can use atmospheric air and extract oxygen through skin, air bladder etc.

2.2.1. Bony fish

In bony fish, the gas exchange organs are called gills. The gills lie in a branchial chamber covered by a bony operculum. The majority of bony fish species have **five pairs of gills.**

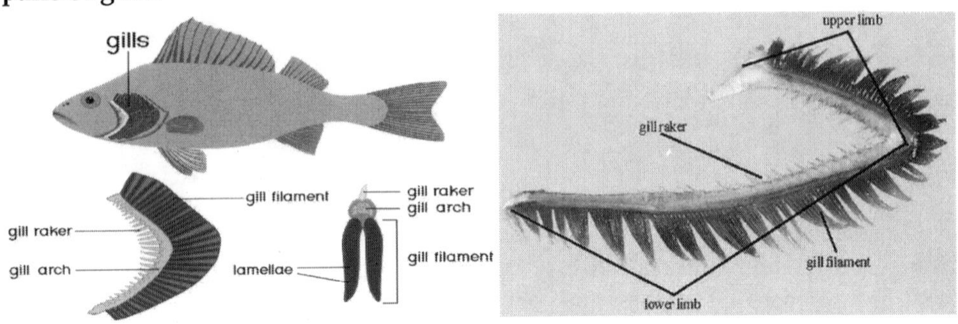

Fig 1.2 – Gills of Bony Fish

The gill arch, along with the filaments, is called a gill. Each gill is composed of many filaments, each covered by many lamellae. The **lamellae** are very fine structures that act as the sites **of gas exchange.** The lamella contains blood capillaries, which have blood flowing in the opposite direction to the water. This is crucial for getting all the available oxygen out of the water and into the blood. This is called the **"counter-current"** mechanism and helps in the maximum exchange of gases, i.e., oxygen intake and carbon dioxide release. If the blood flows in the same direction as the water passing it, it would only be able to get half of the available oxygen from the water. The blood and water would reach equilibrium in oxygen content, and diffusion would no longer occur. By having the blood flow in the opposite direction, the gradient is always such that the water has more available oxygen than the blood, and oxygen diffusion continues to occur after the blood has acquired more than 50% of the water's oxygen content. The counter-current exchange system gives fish **80-90% efficiency** in acquiring oxygen.

Oxygen passes from the water into the blood at the gills. Removal of carbon dioxide also occurs, as the blood containing high concentrations of the waste gas goes to the gills, and the carbon dioxide diffuses out into the water down a diffusion gradient (external water has lower concentrations of carbon dioxide than levels in the blood, so this sets up a diffusion gradient).

2.2.2. Specialized cells in the gills

1. Mucous cells
2. Acidophil granular cells
3. Basophil mast cells
4. Glandular cells
5. Chloride cells

2.2.3. Gill area and diffusion capacity

The gill area of a fish depends on the number and size of the gill filaments and gill lamellae. It has been shown that fast-swimming and active fishes possess more gill area and a higher number of gill lamellae/mm of gill filament than sedentary species. Air-breathing fishes mainly have **accessory respiratory organs,** and the gill area is reported to be ½ of the area of fishes that depends on aquatic respiration. The total gill area increases with the increase in body weight as the fish grows. The efficiency of the gill as a respiratory organ depends on its oxygen diffusion capacity in water, which is inversely proportional to the blood-water barrier (diffusion distance) i.e., the barrier through which the exchange of gases takes place between blood and water. This barrier is formed by the lamellar epithelium, basement membrane and the flange of the pillar cells. (The pillar cells or pilaster cells are a characteristic feature of the teleost gill and serve as pillars to keep the 2 epithelial layers of the lamella separate and prevent them from collapsing). The barrier thickness is less in species depending on water for respiration than in air-breathing fishes. The teleostean gill can diffuse **60-80% of oxygen** from water to blood at the gills.

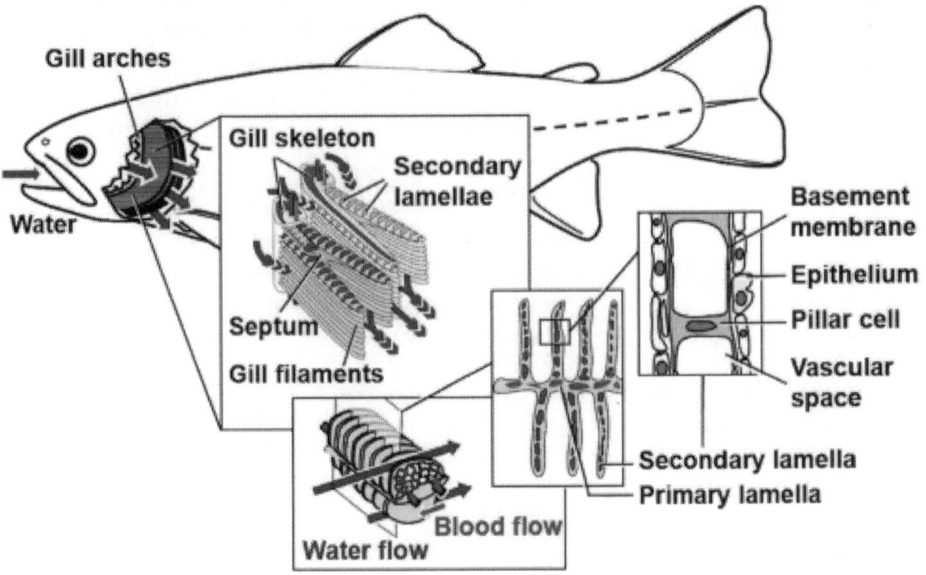

Fig 1.3 – Primary And Secondary Lamella

2.2.4. Mechanism of Respiration - 2 types

- Buccal pumping
- Ram ventilation

Buccal pumping

This is a method of respiration in which the animal moves the floor of the mouth in a rhythmic manner that is externally apparent.

Ram ventilation

The fishes, like **tuna,** swim at high speeds with their mouth parts kept open. There are no visible breathing movements, and water flows continuously over the gills, called ram ventilation. Many fish breathe by pumping at low speeds and change to ram ventilation at higher speeds.

2.2.5. Mechanism of Respiration or Ventilation of gills

For respiration, a continuous water current is maintained through the buccopharyngeal cavity. For this, water is sucked into the buccal cavity and expelled through the external branchial aperture. The inhalation and exhalation of water in the buccal cavity causes blood purification.

When they are moving slowly or resting in water, fish can breathe by synchronously expanding and contracting the buccal cavity (the mouth and throat) and the opercular cavity or an elaborate system of passing water over the gills while the fish stays stationary. In most teleosts, 2 respiratory pumps, buccal and opercular, function in coordination to propel water through the gills. These 2 pumps are operated by a set of muscles resulting in contraction or expansion of the buccal and opercular cavities. As a result, there is an increase or decrease in the pressure inside the 2 cavities so that water flows unidirectionally across the gill lamellae.

The process can be described as below.

1. The mouth is opened, and the buccal cavity is enlarged by lateral expansion of its walls. Enlargement of the buccal cavity lowers the water pressure in it so that water is sucked in.
2. The mouth is closed and the volume of the opercular cavity is increased. However, the opercular opening remains closed so that low pressure is created in the opercular cavity and water flows from the buccal to the opercular cavity through the gill slits.
3. When the operculum is closed, the gills must extract as much oxygen as possible. The blood (high CO_2, low O_2) comes into contact with the water (low CO_2, high O_2) through the microscopically thin membrane of the lamellae. Carbon dioxide seeps into the water and oxygen seeps into the blood vessels. After this, water passes outside due to external branchial aperture through a process called diffusion.

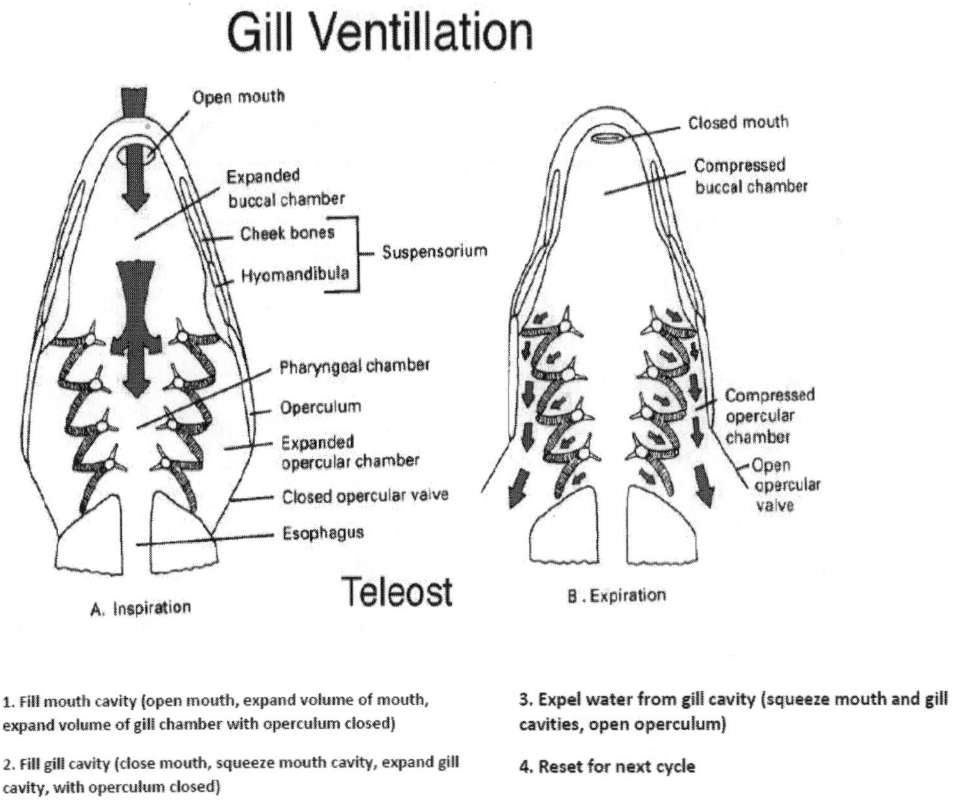

Fig 1.4 – Gill Ventilation

Many sharks rest regularly on the bottom, although they must constantly inhale water to provide the gills with a constant supply. These resting sharks (as well as rays) have structures called spiracles behind their eyes. These holes in the head lead to the gills, supplementing the water taken in by the mouth.

2.3. Fish blood as a gas carrier

Fish blood consists of blood cells and plasma, similar to other vertebrate animals and the blood cells are composed of erythrocytes, leukocytes and thrombocytes. Erythrocytes (red blood cells) are round or ellipse-shaped cells with dense haemoglobin.

Hemoglobin (also spelled haemoglobin and abbreviated Hb or Hgb) has **iron-containing oxygen transport** and Hg is the metalloprotein in the red blood cells of all vertebrates (except the fish family Channichthyidae) and the tissues of some invertebrates. Hemoglobin in the blood transports oxygen from the gills to the rest of the body (i.e., the tissues), where it releases the oxygen for cell use. Haemoglobin is a large protein with four polypeptide chains and four heme groups. Each heme group has an iron atom attached at the centre from the pigment called heme, which gives blood its red colour and its ability to combine with oxygen to be carried to cells and tissues. Each heme group is enfolded in one of 2 or 4 chains of amino acid units that collectively constitute the protein part of the molecule called globin. The haemoglobin of most vertebrates has molecular weights of approximately 65,000 daltons.

Crocodile icefish, or white-blooded fish from the Channichthyidae family, is a perciform fish in the cold waters around Antarctica and southern South America. Their blood is transparent because it contains less than 10% hemoglobin and only defunct erythrocytes. Oxygen is absorbed directly from the water through their scaleless skin. It is then dissolved in the plasma and transported throughout the body without the hemoglobin protein. The loss of hemoglobin is not fatal because of the cold environment in which channichthyids live. Cold water has much higher dissolved oxygen content than warmer water.

2.4. Counter Flow Mechanism

The stream of **water** flowing over the gill and the stream of **blood** running within the gill flow in **opposite directions** to each other, and this helps effectively remove oxygen, which is called counter-current flow. Just as blood is about to leave the gill lamellae, it encounters water whose oxygen has not yet been removed. Thus, this blood takes up oxygen from water, which still has the full oxygen content of inhaled water, and this permits the O_2 content of the blood to reach the highest possible level. As the water runs further between the lamellae, it meets blood with lower and lower oxygen content, and it, therefore, continues to give up more oxygen. The lamellae, along its entire length, serves in taking up oxygen from water, and the water may leave the gill having lost as much as 80 - 90% of its initial oxygen content.

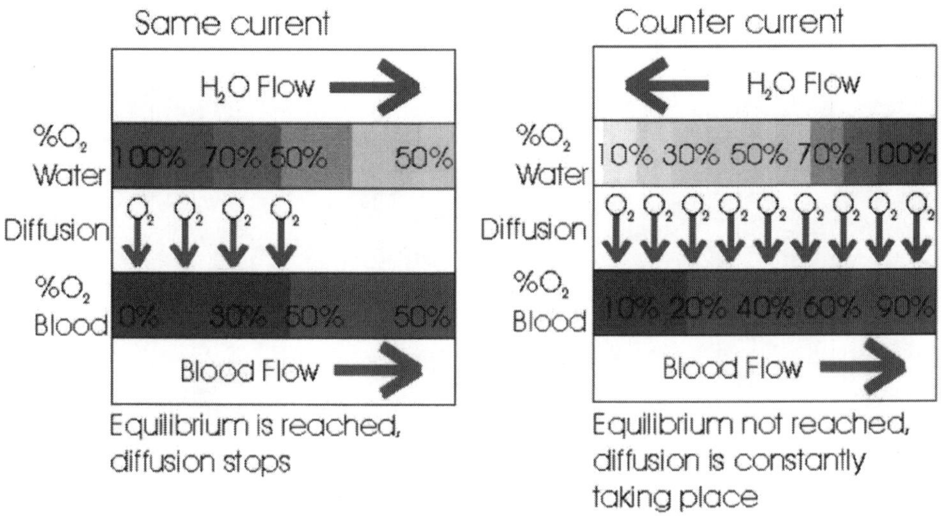

Fig 1.5 – Counter Flow Mechanism

2.5. Oxygen dissociation curve

At high oxygen concentration, the hemoglobin (Hb) combines with oxygen to form oxyhemoglobin (HbO_2)

$4Hb + 4\,O_2 \rightarrow 4HbO_2$

At low concentrations, oxygen is given up again and the reaction proceeds to the left.

The haemoglobin is saturated according to the given O_2 concentration. At any given oxygen concentration, there is a definite proportion between the amount of haemoglobin and oxyhaemoglobin. Thus, a oxygen haemoglobin dissociation curve is obtained by plotting the oxyhaemoglobin for a given oxygen concentration. This curve will be generally in sigmoid shape.

In some animals, the blood has a high oxygen affinity (it gives up oxygen less readily), and the dissociation curve is located further to the left. The blood of others gives up oxygen more readily (the oxygen affinity is lower), and the dissociation curve is located further to the right. The binding of oxygen to haemoglobin is affected by several factors.

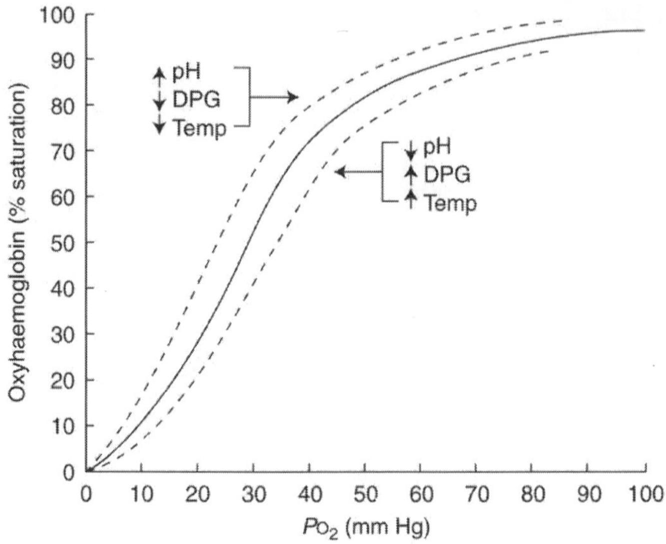

Fig 1.6 - Oxygen dissociation curve

2.5.1. Effect of temperature, CO_2 and pH

An increase in temperature weakens the bond between haemoglobin and oxygen and causes an increased dissociation of the bond. So, the blood gives oxygen more readily and the dissociation curve is shifted to the right. This helps to deliver oxygen at higher temperatures readily in cold-blooded animals.

An increase in CO_2 or other acids lowers the pH of the plasma, shifting the dissociates curve to the right. This effect is called the **Bohr** effect or **Bohr shift**.

2.5.2. Bohr effect

The Bohr effect involves a decreased affinity of haemoglobin for oxygen due to the offered configuration of haemoglobin molecule by binding with Hydrogen ion (The oxygen dissociation curve is shifted to the right). These arrangements lead to oxygen loading at respiring tissues, where CO_2 is high. This CO_2 rapidly goes out at gills, and it dissociates from haemoglobin, allowing for the effective loading of oxygen.

2.5.3. Root effect

The Root effect is an extension of the Bohr Effect, mainly in fishes with gas bladders. This effect involves a decrease in the oxygen capacity of the blood with a rising partial pressure of CO_2 (or otherwise reduction is pH). A decrease in pH will render fish incapable of becoming 100% saturated with oxygen regardless of the pressure of oxygen and this facilitates the effective transfer of O_2 from haemoglobin.

This phenomenon is explained by the fact that fish blood contains two different haemoglobin species: one is highly sensitive to acid (anodic Hb), and the other is acid-insensitive (cathodic Hb). In the presence of high carbon dioxide, the former carries no oxygen, while the letter presents the usual dissociation curve for this haemoglobin species. Fishes produce lactic acid during high activity. The presence of lactic acid in the blood made all haemoglobin incapable of binding oxygen, and the fish would die of asphyxia. However, the presence of acid-insensitive haemoglobin avoids this calamity.

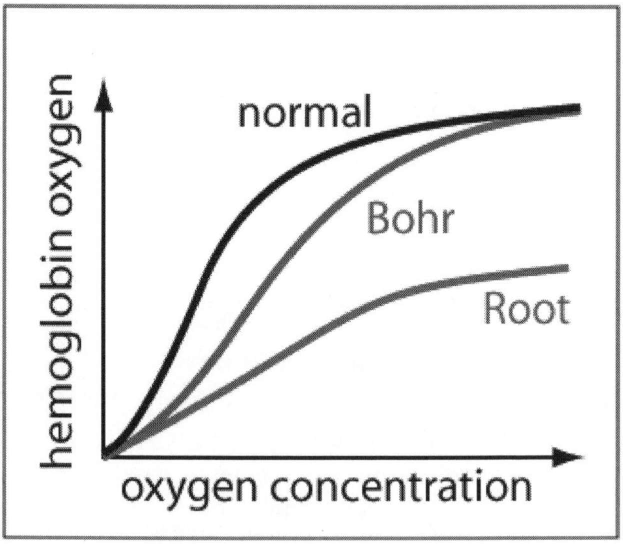

Fig 1.7 – Bohr and Root effect

2.5.4. Effect of organic phosphates

The presence of organic phosphates, adenosine triphosphate or guanosine triphosphate or 2, 3 disphosphoglycerate in the blood cells reduces the oxygen affinity by haemoglobin and a shift to the right of the oxygen dissociation curve is noticed.

2.6. Carbon dioxide transported in the blood

The transportation of carbon dioxide is a very significant process in the gas-transfer systems of many animals. There are three main ways in which CO_2 is transported in the blood. A small percentage of the CO_2 in the blood is dissolved as molecular CO_2. A more significant amount of CO_2 reacts with $-NH_2$ groups of haemoglobin and other proteins to form carbamino compounds. However, most of the CO_2 in the blood is transported in the form of bicarbonate (HCO_3^-). The change of bicarbonates into carbon dioxide and water is catalyzed by the enzyme carbonic anhydrase. This enzyme is present in the acidophil cells of the gills, red blood cells and other tissues but is found absent in the plasma.

$$CO_2 + H_2O \overset{CA}{\leftrightarrow} H^+ + HCO_3^-.$$

In general, CO_2 is diffused into the blood from the tissues. The blood transports CO_2 to the respiratory surfaces of the gills, where it is released into the environment. The blood mainly consists of plasma and erythrocytes (red blood cells). Most of the CO_2 entering and leaving the blood does through erythrocytes.

Mackerel is a fish of high seas that never encounters low oxygen levels and lives under uniform but low carbon dioxide tensions. The mackerel blood is very susceptible to small changes in carbon dioxide, whereas the blood of the carp and the bullhead catfishes, who often live in stagnant waters, is relatively unaffected by changes in carbon dioxide levels.

Factors that influence oxygen consumption include activity, age, reproductive state, nutrition, disease and intrinsic regulatory mechanisms (nervous and hormonal control).

Increases in carbon dioxide tensions or decreases in pH lead to decreased oxygen consumption. Carbon dioxide is considerably more soluble in water than oxygen.

2.7. Accessory respiratory organs in fishes

Accessory respiratory organs in fish are generally found in the fishes that live in shallow, stagnant freshwater in tropical regions where deoxygenation of water is a prominent feature. Such structures may also develop in fishes inhabiting streams of the hills, which are liable to dry up during summer.

2.7.1. Buccopharyngeal epithelium

This is the most primitive type of adaptation and has been found in *Monopterus javanensis, Electrophorus electricus, Periopthalamus* sp and *Bleopthalamus* sp. In these fishes, the epithelial lining of the mouth cavity is highly vascular and forms a vascular network, thus making it an efficient respiratory site.

2.7.2. Skin

The extent to which fishes use cutaneous respiration is poorly known. It is best documented in eels (*Anguilla anguilla*), which leave the water and migrate from one place to another. During this time, the moist skin (richly supplied with blood) works as an essential organ for aerial respiration. The skin is of little use for respiration in some fishes like Monopterus and Mastacembelus, which live in oxygen-deficient stagnant waters. The glandular secretions of the skin protect skin from desiccation in water-deficient conditions.

2.7.3. External gills

These are highly vascular and filamentous structures of the ectoderm covering the outer surface of the visceral arches or visceral skeleton. They are in direct contact with water, so they help in the exchange of gases. To aid respiration, the embryos of certain elasmobranchs possess long filamentous structures present as outgrowth from spiracles. (External gills - develop from the skin ectoderm of the branchial area but are not directly related to the visceral skeleton or branchial chambers and are found most often in larval or paedomorphic amphibians)

2.7.3. Labyrinthiform organs

These organs have been present in a few perches like *Anabas testudineus* and *Trichogaster fasciatus*. These are formed within the extrabranchial chambers enclosed between the gills and the operculum. These organs originate from the 1st gill arch.

2.7.4. Opercular lungs

In advanced air-breathing teleosts, the bag-like diverticulae are called opercular lungs or air chambers. It develops from the dorsal surface of the branchial chamber or opercular chamber. The opercular lungs are above the gills and contain specialized structures known as rosettes or arborescent organs (arbo = tree), which increase the respiratory surface. Such accessory respiratory organs are found in *Clarias, H. fossilis, A. testudineus, T. fasciatus* etc.

2.7.5. Air bladder/ swim bladder

The swim bladder is present between the gut and the kidney. This duct degenerates in the course of development or is present throughout life. The fishes with air bladders have adapted to extensive drought conditions, which may cause them to dry up completely. The teleosts like *Chirocentrus,* and Cyprinoids come to the surface, take air and pass it back to the air bladder, which is highly vascular. The swim bladder is a gas-filled, bag-like structure that originates from the dorsal outgrowth of the alimentary canal in most ray-finned fishes. It controls buoyancy at different depths in water bodies. The gases in the air bladder consist primarily of oxygen, carbon dioxide and nitrogen.

2.7.6. A special part of the alimentary canal

In small groups of fishes, inhaled air is swallowed and forced back inside the alimentary canal, a part of which may be modified for aerial respiration. After respiratory exchange, the used air is either passed outside by the anus or is expelled through the mouth. Some modifications have occurred in the alimentary canal to help with gaseous exchange. There is an excellent reduction in stomach and intestine muscle layers, making it thinner and transparent. The inner surface is lined by a single layer of epithelial cells, which is highly supplied with blood.

2.8. Respiration in Cartilaginous fish

Sharks and rays typically have **five** pairs of slits that open directly to the outside of the body, though some more primitive sharks have six or seven pairs, and they don't have operculum. Gas exchange occurs at the gills, and oxygenated water must always flow over the gill filaments for respiration to occur. Water enters through the mouth of the animal, into the pharynx, over the gills and exits through the gill slits. Respiratory gas exchange occurs on the gill filaments' surface as the water passes over and out the gills.

Adjacent slits are separated by a cartilaginous gill arch from which projects a long sheet-like septum, partly supported by a further piece of cartilage called the **gill ray**. The individual lamellae of the gills lie on either side of the septum. The base of the arch may also support gill rakers, small projecting elements that help to filter food from the water.

2.8.1. Spiracle

A smaller opening, the spiracle, lies in front of the first gill slit. This bears a small **pseudobranch** resembling a gill in structure but only receives blood oxygenated by the true gills.

The spiracle is a vestigial first gill slit. It appears as an opening **behind the eye**, as in the spiny dogfish. It is absent or reduced in many sharks, especially the fast-swimming sharks and is usually more prominent and present in sedentary or bottom-dwelling sharks. The spiracle in sharks is used to provide oxygenated blood directly to the eye and brain through a separate blood vessel. In the rays, the spiracle is much larger and more developed and is used to actively pump water over the gills to allow the ray to breathe while buried in the sand.

Most sharks rely on ram ventilation, forcing water into the mouth and over the gills by rapidly swimming forward. In slow-moving or bottom-dwelling species, especially among skates and rays, the spiracle may be enlarged, and the fish breathes by sucking water through this opening instead of through the mouth.

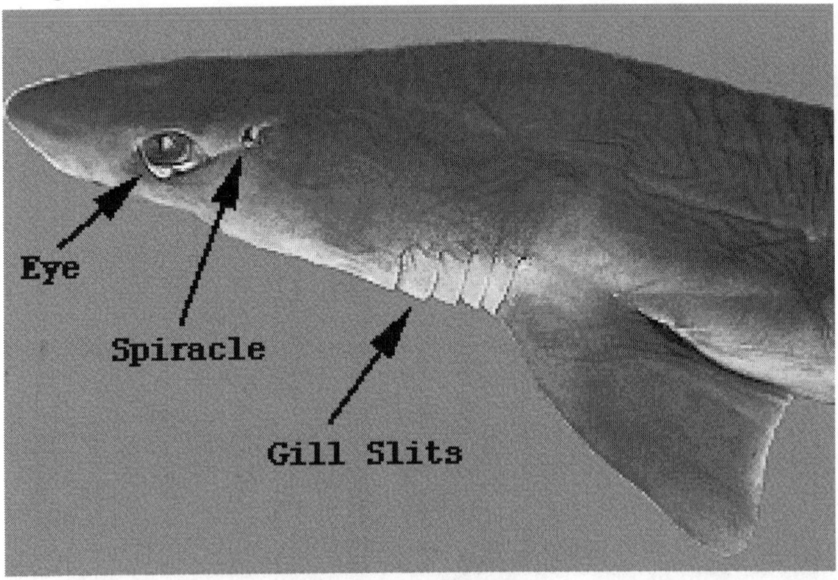

Fig 1.8 - Spiracle

2.9. Respiration in shellfish

2.9.1. Haemocyanin - Respiratory Pigment in Shellfish

Haemocyanin is the respiratory pigment in most arthropods and molluscs. It is a high molecular weight protein (75 kD) containing two copper (Cu) atoms. It is dissolved in blood plasma. Deoxygenated blood is white, while oxygenated blood is blue in colour. Normal stability of oxygen binding is achieved only when the unit contains 8 copper atoms (tetramer). Oxygen-carrying capacity is low compared to haemoglobin.

Shellfish is any aquatic invertebrate animal with a shell belonging to the phylum Mollusca, Arthropoda or Echinodermata. Bivalve molluscs, including oysters, mussels, scallops, and clams. Gastropod molluscs, such as abalone, whelk, and conch. Arthropoda includes shrimp, lobster and crabs. Echinodermata includes starfish, sea urchins and sea cucumber.

2.9.2. Respiratory system in shrimp

The respiratory system in shrimp consists of three principal organs.

1. **Lining of branchiostegites or gill covers:** It is a thin, membranous and highly vascular lining containing minute blood lacunae or vessels. This organ forms a large surface area for the absorption of O_2 and release of CO_2 in water.
2. **Epipodites:** These are a simple, foliaceous and highly vascular outgrowth of the integument that arises from coxal segments of maxillipeds. These epipodites in the anterior part of the gill chamber carry out respiratory functions.
3. **Gills**: These are the primary respiratory organs. There are eight gills inside each gill chamber, each attached to the thoracic wall by a gill-root. These gills are crescent-shaped and their size increases gradually from anterior to posterior. Shrimp gills are of **phyllobranch** type**, each consisting of two rows of leaf-**like gill plates. Gill plates are arranged like the leaves of a book. Gill plates are the largest in the middle but become gradually smaller towards the end. A deep median longitudinal groove runs between the two rows of gill plates.

Types of gills according to their place of origin and attachment

1. **Podobranch or foot gill:** It is attached to the coxa of each second maxilliped.
2. **Anthrobranch or joint gill:** It is attached to the arthrodial membrane, joining a limb with the body. Each third maxilliped bears two arthrobranchs.
3. **Pleurobranch or side gill:** It is attached to the lateral wall of the thoracic segments bearing the 5 walking legs

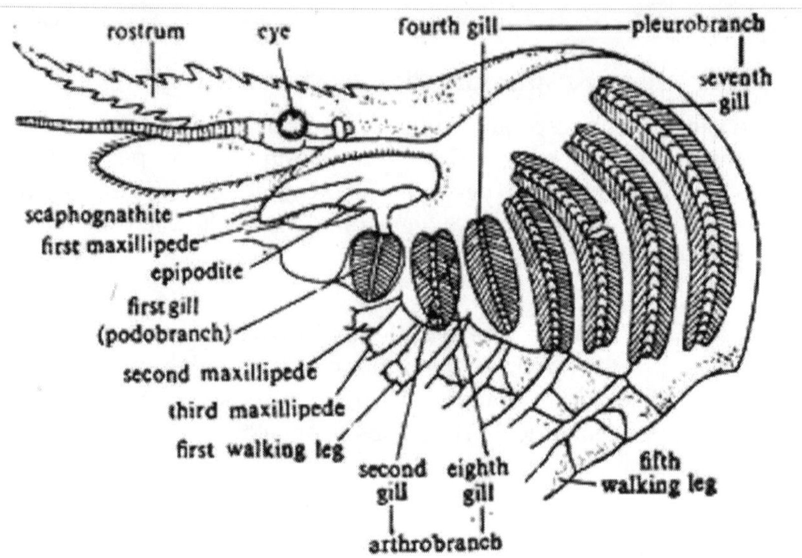

Fig 1.9 – Respiratory organs of *Palaemon*

2.9.3. Blood Supply

Three longitudinal blood channels run through the gill base from one end to the other. Two are lateral longitudinal and the third is a median longitudinal channel running through the gill base, beneath the median group of gills. In each gill plate, the lateral channel gives off a slender marginal channel, which finally joins the median longitudinal channel.

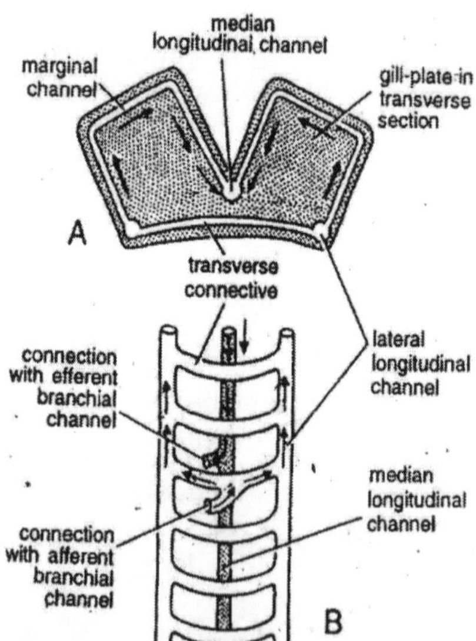

Fig 1.10 - Blood Supply

2.9.4. Blood Circulation in Gill

Deoxygenated blood from the body is brought to the gill by an afferent branchial channel. It enters through the gill root, flowing first through the lateral longitudinal channels, then marginal channels and finally reaches the median longitudinal channels. During this path, blood gets oxygenated. An efferent branchial channel carries blood from the median channel to the heart.

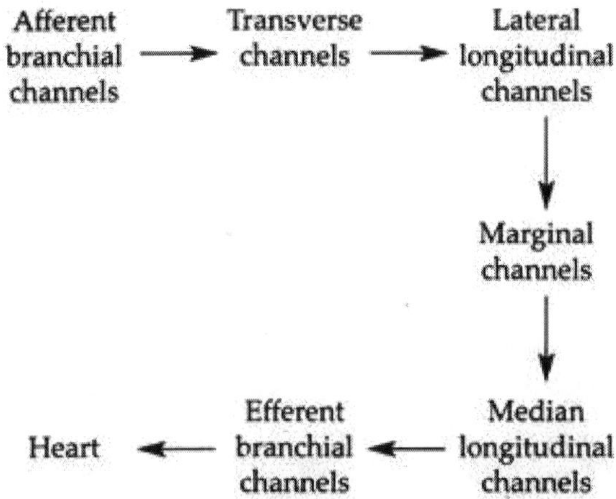

Fig 1.11 - Blood Circulation in Gill

2.9.5. Mechanism of Respiration in Prawn:

As a result of the constant anteroposterior movement of the exopodite or scaphognathite of the second maxilla, a water current enters the gill chamber through the posterior side. Movements of exopodites of maxillipeds force the water to rush inside the gill-chamber through the posterior and lateral sides.

During the flow of water currents, the vascularized surface of branchiostegites, gills and epipodites are bathed in water. The gaseous exchange occurs through these areas as the dissolved oxygen is taken in and carbon-dioxide passes out. The carbon-dioxide mixed water is expelled out through the ventral region of the gill-chamber due to the movement of scaphognathite and epipodites. In prawn, the respiratory pigment, haemocyanin, is dissolved in plasma and carries oxygen to the tissue cells.

2.9.6. Respiratory system and gas exchange in lobster

The gills of lobsters are known as **trichobranch** type. Twenty pairs of gills within the branchial chambers on either side of the cephalothorax comprise the respiratory system. These gills are made up of numerous feathery-like filaments situated around a central rod and are protected within the gill chamber of the carapace. Water passes through openings between the lobster legs, over the gills and up towards the head. This water current is reversed every few minutes so debris can be flushed out of the chambers. On any given thoracic segment, there may be as many as four pairs of gills: one pair on the basal segment of the limbs, two pairs arising from the soft membrane linking the limbs to the body, and one pair on the side of the body just above the limbs. In the American lobster, the full complement of gills occurs at the base of the second, third, and fourth thoracic legs, with fewer pairs on the remaining thoracic segments except the first, which lacks gills.

On each side, the gills lie within the branchial chamber, formed by a deep lateral fold of the carapace. Access to the brachial chamber is through tiny openings between the appendages and two larger openings, both ventral, one at the posterior end of the branchial chamber and the other at its anterior end. At the anterior opening, there is the leaf-like flap known as the gill bailer, or scaphognathite, which performs rapid beating and drives water forward in the channel and out of the branchial chamber.

2.9.7. Respiratory system of the true crab

The primary respiratory organs of true crabs are called gills, which are arranged like a pyramid within the two gill-chambers. Each gill chamber is on the side of the thoracic chamber and completely separated from it by pleuron. Externally, the gill chamber is covered by the branchiostegite. Gill of crabs is also known as phyllobranch.

Water enters the branchial chambers of crabs primarily through an anterior opening above the base of each claw, or cheliped, and to a much lesser extent through openings at the base of the other thoracic legs. In the blue crab, when the chelipeds are raised and held forward, the opening at the base of the chelipeds is huge and nearly circular. When the chelipeds are folded against the body, the opening is a wide slit. Bristle-like setae arising from the basal portion of the cheliped filter the water entering the slit. The current of water through the branchial chambers of crabs is maintained by the beating of the gill bailer, which lies in the channel at the anterior exhalant opening of each branchial chamber. The water passes under the gills, up between the gills, over the gills, and out through the exhalant aperture.

Crabs have articulating plates for breathing out of the water that can keep their gills moist by sealing them in.

2.9.8. Respiration in Mollusca (Pila)

The respiratory organs consist of a single gill or ctenidium, (2) a pulmonary sac or lung and (3) a pair of nuchal lobes.

i) Ctenidium:

It consists of numerous thin lamellae attached to the mantle wall by their broad base and apices free in the branchial chamber. A ctenidium with a single row of lamellae attached to an axis is called monopectinate gills. The lamellae present in the middle of the ctenidium are largest and they gradually become smaller towards the ends. The epithelial layer of lamellae consists of three kinds of cells: non-ciliated columnar cells, ciliated columnar cells and glandular cells.

ii) Pulmonary Sac:

It is a sac-like structure hanging from the roof of the mantle cavity. It is developed from the mantle and consists of a densely pigmented dorsal wall and a creamy white ventral wall. The flaps on two sides of the openings are of unequal size, the left being larger than the right, and they help control the opening of the pulmonary sac.

iii) Nuchal Lobes:

The mantle has two fleshy and contractile nuchal lobes on the two sides of the head. The left nuchal is well developed, whereas the right is less developed. They form longer and smaller respiratory tubular funnels during respiration.

2.9.8.1. Mechanism of Respiration

Aquatic respiration:

The animal is fully expanded, and the left nuchal lobe increases in size and takes the form of a gutter. The respiratory current of water enters through the gutter-like left lobe and reaches the posterior part of the pulmonary chamber. Later, it reaches the branchial chamber, bathing the entire length of the gills.

Aerial or pulmonary respiration:

Pila moves towards the surface of the water and thrusts out its left nuchal lobe, which becomes elongated and tubular and forms a respiratory siphon. The pulmonary chamber becomes greatly enlarged and cut off from the branchial chamber. Air enters the pulmonary sac through the respiratory siphon, and respiratory movements produced by the alternate contraction and expansion of the pulmonary sac help in respiration. The respiratory siphon is not formed in terrestrial animals, and the air enters their pulmonary chamber directly through the expanded nuchal lobe.

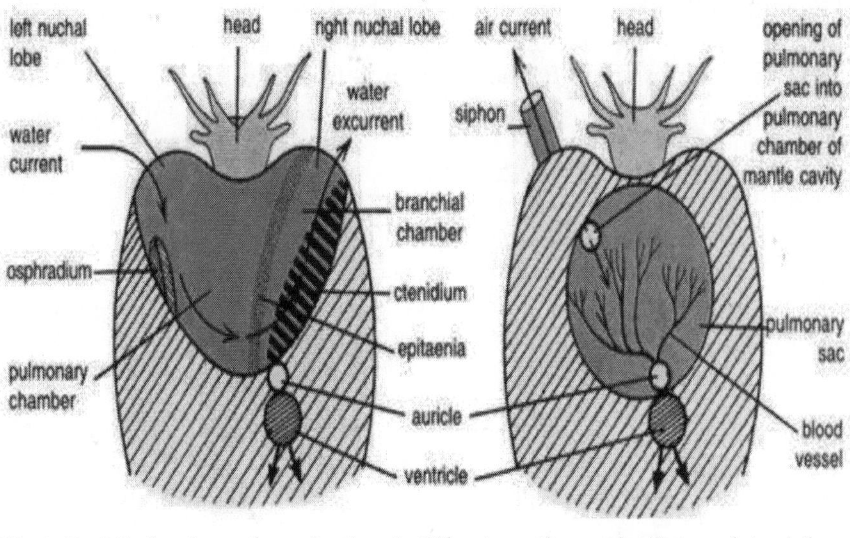

Fig 1.12 - Mechanism of respiration in Pila. Aquatic respiration and Aerial respiration

2.9.9. Respiratory System of Unio

Respiration is aquatic and carried on simultaneously with the feeding process. The respiratory organs are the gills or ctenidium and the mantle. The freshwater muscle respires the oxygen dissolved in water by a pair of gills or ctenidia or branchiae. A single gill on each side of the foot, hanging the mantle cavity between the mantle and the visceral mass like a flattened, plate-like structure.

There is a single pair of elongated ctenidia or gills on each side of the foot, which are of the eulamellibranch type. Each ctenidium appears double, but it is made of two gill plates. These are called demibranchs or laminae, containing an outer and an inner gill plate derived by the folding of a single ctenidium. Each ctenidium is bipectinate, with a central ctenidial axis, from which filaments arise in two rows, one on either side. The filaments of each row are folded in the middle to appear V-shaped in section. Of the two arms of V, one is descending, the other ascending. At the angle of the fold, each filament is notched. The notches of all filaments form a continuous food groove that extends the whole length of the underside of each lamina of the ctenidium.

The adjacent gill filaments are joined by tissue fusion, forming an interfilamentar junction. Thus, gill filament and interfilamentar junctions form the two lamellae of a gill plate. In the interfilamentar junctions are holes known as ostia, which connect the infrabranchial chamber of the mantle cavity with water tubes in the laminae. The gill filaments are covered with various kinds of cilia; each gill filament is supported by two chitinous rods. The lateral cilia are found on the sides of filaments, where the frontal cilia cover the distal surface and the latero-frontal cilia cover the surface of the lateral side

2.9.9.1. Attachment of Ctenidia

The outer lamella of the outer gill plate is attached to the mantle. The inner lamella of the outer gill plate and the outer lamella of the inner gill plate are joined together to the visceral mass. The inner lamella of inner gill plates is attached to the visceral mass anteriorly, but further back, it is free, and behind the foot, it is joined to its fellow of the other side so that the inner lamellae of inner gill plates are united with one another.

2.9.9.2. Blood Supply of Gills

The ctenidia are supplied by afferent branchial vessels carrying deoxygenated blood from the kidneys and divide to give rise to branches into the interlamellar junctions. These branches unite to open into the efferent branchial vessel, which carries away oxygenated blood to the heart.

2.9.9.3. Course of Water Current

The constant beating of the cilia of gill filaments causes a continuous current of water to enter the inhalant siphon lying posteriorly and ventrally from where it goes to the mantle cavity. The lateral cilia pass water current to water tubes of gill plates through ostia, then it goes to the supra-branchial chamber and passes out through the exhalant siphon. Water currents transport oxygen and bring food, and the outgoing currents carry away excretion and faeces products in addition to CO_2. The mantle also helps in respiration. It is richly supplied with the blood vessels, and gaseous exchange occurs through its thin wall.

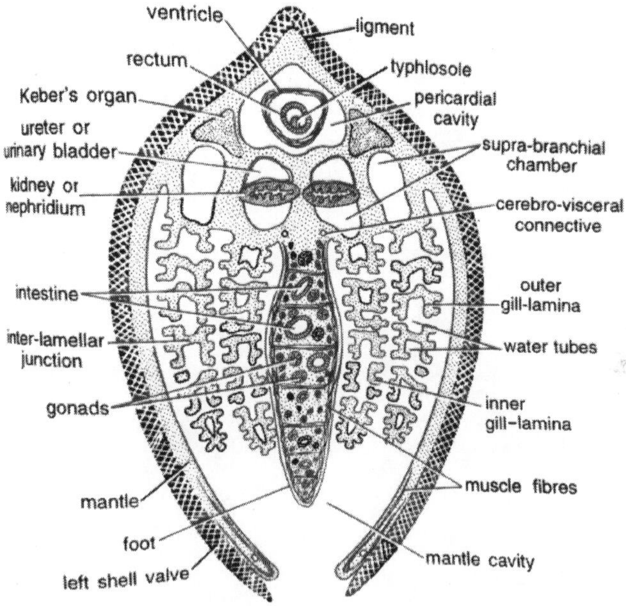

Fig 1.13 - Respiratory System of Unio

CHAPTER 3

MUSCLE PHYSIOLOGY

3.1. Introduction

In higher animals, such as vertebrates, most movements are caused by **skeletal, cardiac and smooth muscles**. The skeletal muscles, which have visible stripes or striations when viewed with the microscope, usually contract rapidly and are under the voluntary control of the animal. The cardiac muscle, which makes up the heart, is unique as it is striated but not under voluntary control. It continues pumping automatically throughout life. Smooth muscle, which does not have microscopic striations, usually contracts very slowly and often undergoes rhythmic pulsations.

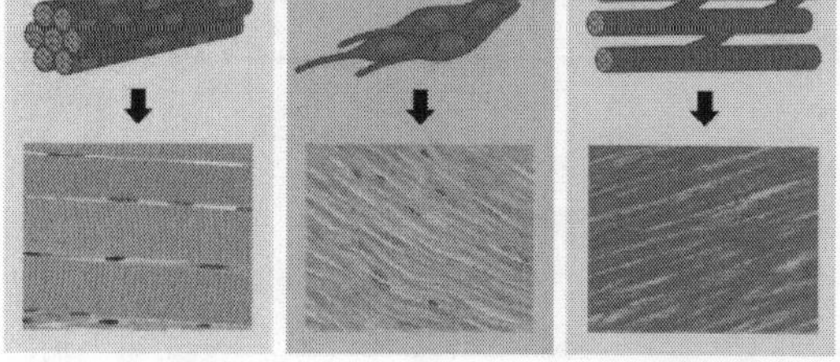

Fig 1.14 - Skeletal, Cardiac and Smooth muscles

3.2. Striated or striped muscle

Striated or striped muscle constitutes a significant fraction of the total body weight. The movement of the limbs depends on the contraction of the striated muscle, and they maintain posture. The ends of most striated muscles are attached to the skeleton and thus are often called skeletal muscles. They are attached to the bones by tendons. Each striated muscle has blood vessels and nerves associated with it. The vessels transport blood to and from the muscle, supplying oxygen and nutrients and removing CO_2 and other wastes.

3.2.1. The muscle fibre

Living muscle comprises many long, cylindrical-shaped fibres of 0.02 to 0.08 millimetres in diameter. In some muscles, the fibres run the entire muscle length, up to several centimetres long. In others, a tendon extends along each edge, and the fibres run diagonally across the muscle between the tendons. A connective tissue sheath surrounds each muscle called the epimysium. Fascia, connective tissue outside the epimysium, surrounds and separates the muscles. Portions of the epimysium project inward to divide the muscle into compartments. Each compartment contains a bundle of muscle fibres. Each muscle fibre bundle is called a fasciculus and is surrounded by a layer of connective tissue called the perimysium. Within the fasciculus, each muscle cell, called a muscle fibre, is surrounded by connective tissue called the endomysium.

The plasma membrane of muscle fibres is called the **sarcolemma**, and the cytoplasm is called the **sarcoplasm**. Within a muscle fibre, proteins are organized into organelles called **myofibrils** that run the length of the cell and contain sarcomeres connected in series. Because myofibrils are only approximately 1.2 μm in diameter, hundreds to thousands (each with thousands of sarcomeres) can be found inside one muscle fibre. The **sarcomere** is the smallest functional unit of skeletal muscle fibre and is a highly organized arrangement of contractile, regulatory, and structural proteins.

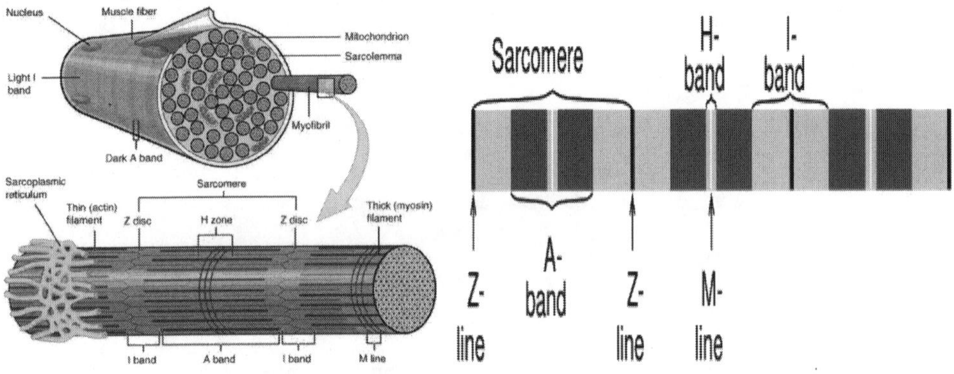

Fig 1.15 - Muscle Fibre

3.2.2. Myofibril

A myofibril or a muscle fibril is a basic rod-like unit of a muscle cell. There are hundreds of myofibrils running parallel to each other in a muscle cell. Myofibrils are composed mainly of actin and myosin proteins. Some other types of proteins are also present in myofibrils. These proteins are organized into thick and thin long filaments called **myofilaments**. Thin myofilaments consist primarily of actin protein, while thick filaments comprise myosin protein. These two types of myofilaments run through the length of the myofibril in sections called sarcomeres. Myofibrils comprise repeating sections of sarcomeres. These sarcomeres appear as alternating dark and light bands under the microscope and are responsible for muscle contractions.

3.3. Cardiac muscle

The heart consists primarily of cardiac muscle cells (or myocardium). Cardiac muscle cells are striated and branched, contain many mitochondria, and are under involuntary control. Each myocyte contains a single, centrally located nucleus surrounded by a cell membrane called the sarcolemma. The sarcolemma of cardiac muscle cells contains voltage-gated calcium channels, specialized ion channels that skeletal muscle does not possess. The outstanding characteristics of the action of the heart are its contractility, which is the basis for its pumping action, and the rhythmicity of the contraction.

The cardiac muscle is responsible for the contractility of the heart and, therefore, the pumping action. The cardiac muscle must contract with enough force and enough blood to supply the metabolic demands of the entire body. This concept is termed cardiac output and is defined as heart rate x stroke volume.

3.3.1. Pacemaker cells of the heart

The contraction of cardiac muscle (heart muscle) in all animals is initiated by electrical impulses known as action potentials in the heart, known as cardiac action potentials. The rate at which these impulses control the rate of cardiac contraction is known as heart rate. The cells that create these rhythmic impulses setting the pace for blood pumping, are called pacemaker cells, which directly control the heart rate. They make up the cardiac pacemaker, that is, the natural pacemaker of the heart.

3.4. Smooth muscle

Smooth muscle (so-named because the cells do not have striations) is present in the walls of hollow organs like the urinary bladder, uterus, stomach, intestines and in the walls of passageways, such as the arteries and veins of the circulatory system, and the tracts of the respiratory, urinary, and reproductive systems.

Smooth muscle fibres are spindle-shaped (wide in the middle and tapered at both ends, somewhat like a football) and have a single nucleus; they range from about 30 to 200 μm (thousands of times shorter than skeletal muscle fibres), and they produce their connective tissue, endomysium. Although they do not have striations and sarcomeres, smooth muscle fibres have actin and myosin contractile proteins.

3.5. Visceral muscle

Visceral muscle is found in the body's internal organs and blood vessels. It is usually called smooth muscle because it has no striations and is, therefore, smooth in appearance. It is found as layers in the mucous membranes of the respiratory and digestive systems.

3.6. Physiology of muscles

The immediate source of energy for muscle contraction is **adenosine triphosphate** (ATP), a compound that is the primary energy source. ATP is the only substance that the muscle proteins can use directly. When the terminal phosphate group of this compound is split off, the high energy of the bond is available for the energy of muscle contraction.

Despite its great importance, ATP is present in muscle in minimal amounts. The total quantity may be sufficient for not more than ten rapid contractions; therefore, ATP must be restored rapidly, for otherwise, the muscle would soon be exhausted. The source is another organic phosphate compound, creatine phosphate, which is present in more significant amounts. Its phosphate group is transferred to adenosine diphosphate (ADP), thus restoring the supply of ATP.

The creatine phosphate must, however, eventually be replenished, and the ultimate energy source is the oxidation of carbohydrates. Carbohydrate is stored in the muscle in the form of glycogen, which is often present in an amount of between 0.5 and 2% of the wet weight of the muscle and provides an amount of energy perhaps 100 times as significant as the total quantity of creatine phosphate. Without sufficient oxygen, the glycogen can still yield energy by being split into lactic acid, but the amount of energy is only a tiny fraction.

Contraction of the muscle causes a force to be exerted on the points of attachment; because no change in length takes place, this is called an *isometric contraction*. If, on the other hand, we attach a weight to one end of the muscle, it can lift, and the muscle shortens during contraction because the load remains the same throughout the contraction; this is called an *isotonic contraction*.

The molecular events during muscle contraction are better known than most other processes in the living organism. To describe these events, we must refer to the structural composition of the thick and thin filaments.

The **thick** filaments consist of the protein ***myosin***. Each myosin molecular resembles a thin rod with a globular "head". One thick filament contains several hundred myosin molecules lined up with the heads facing in both directions away from the centre of the filament. This leaves a bare zone in the middle, and the heads protrude throughout the thick filament.

The **thin** filaments are more complex, comprising three crucial proteins.

Actin, a relatively small globular protein, is arranged in the filament as a twisted double strand of beads. Each actin molecule is slightly asymmetrical, and in each filament, all the actin is lined up and facing in the same direction. This directionality is vital for muscle contraction.

The next important protein in the thin filament is ***tropomyosin***, long and thin molecules attached end to end, forming a fragile threadlike structure in the grooves between the double helix of the actin molecules. The length of one tropomyosin molecule is such that it extends over seven actin molecules. The single thin filament is about 1 μm long and contains about 400 actin molecules and 60 tropomyosin molecules.

A third protein molecule, ***troponin***, is attached to each tropomyosin molecule. Troponin, a calcium-binding protein, is a key to the contraction process. When troponin binds calcium ions, it undergoes a conformational change that is essential for the interaction between the myosin heads of the thick filaments and the actin of the thin filaments.

The interaction between thick and thin filaments consists of a cyclic attachment and detachment of cross-bridges between the two filaments. The myosin heads attach to the thin filaments at a certain angle; they undergo a conformational change that makes the bridge swivel to a different angle, pulling the thin filament past the thick. The cross bridges on opposite sides of the bare zone in the middle of the thick filament swivel in opposite directions, pulling at the opposing ends of the thin filaments, thereby shortening the distance between the Z-lines. This decrease in the distance between the Z-lines causes the shortening of the muscle.

3.7. Initiation of muscle contraction

Normally, muscle contraction is initiated when a nerve impulse arrives at the neuromuscular junction *(the motor endplate)*. The impulse spreads rapidly as an electric depolarization over the surface of the muscle fibre (the *sarcolemma)*, momentarily abolishing the normal surface potential of about – 60 mV. This depolarization is almost instantaneously communicated to make contraction take place.

The muscles of fish also have twitch and tonic fibres. Pelagic fish such as mackerel and tuna swim continuously at relatively low speeds and have two types of fibres, separated in different muscle masses with strikingly different appearances. The tonic muscle is deep red because of a high concentration of myoglobin. It is located along the sideline and stretches toward the vertebral column. This red muscle entirely executes the basal swimming during cruising; the large mass of white muscle (twitch type) represents a reserve of power during prolonged periods of slow swimming. However, there is no change in their glycogen level, indicating that they are not used during cruising.

The separation into red and white swimming muscles has an interesting counterpart in an invertebrate. In squid, the mantle muscles show a clear separation between layers with a high density of mitochondria and oxidative enzymes and a layer low in mitochondria but high in glycolytic enzymes. This metabolic differentiation is analogous to the red muscle of fish that supports steady swimming and the white muscle that supports burst-type swimming.

3.8. Shellfish Muscle Physiology

3.8.1. Molluscs muscles

Smooth muscles in Molluscs are called "catch" muscle fibres, which hold the shells of lamellibranch tightly closed for long periods. They are found in the opaque (or white) part of the adductor and in the byssus retractor muscles. In the intact adductor, they can be distinguished readily from translucent and cross-striated fibres by their white, tendon-like appearance.

Clams and mussels protect themselves by closing their shells. One or more adductor muscles pull the shells together and keep them closed against the springy action of the elastic hinge. Species that live in the intertidal zone have elastic hinges. Species living in the intertidal zone must keep their shells closed when the water recedes, and the muscle is contracted for hours with the shells tightly closed under conditions that are probably entirely anaerobic. If a starfish attacks a mussel, it attaches its tube feet to the two shells and tries to pull them apart. During that time, the closing muscle of the clam remains contracted, known as **catch**, which is quite different from an ordinary contraction. A muscle in catch seems to maintain contraction without fatigue and with very little or no increase in oxygen consumption.

In some species, but not all, the closing muscles are divided into two portions, one of **smooth** and the other **striated** fibres. Functionally, the striated portion contracts quickly and is referred to as the fast or phasic (twitch) portion; the nonstriated is much slower and is known as the slow or tonic portion. The long-maintained contractions that keep the shells closed for extended periods are produced primarily by the slow portion of the closing muscle.

The most studied muscle of the catch type is the **anterior byssus retractor muscle (ABRM)** of the blue mussel (*Mytilus edulis*), which contains only smooth fibres. It can be made to contract by electric stimuli, but depending on the nature of the stimulus, relaxation may be complete within a matter of seconds, or contraction may be maintained for minutes or hours.

The nerve to the ABRM of *Mytilus* is a mixed nerve; that is, it contains excitatory and relaxing nerve fibres. The excitatory fibres presumably act through the release of acetylcholine, and the relaxing fibres by the release of another transmitter substance, serotonin. Direct application of serotonin to a muscle in catch causes immediate relaxation. In the absence of an activation of the relaxing system (i.e., in the absence of serotonin), probably all excitatory stimuli cause catch.

The unique feature of the catch is that stiffness of the muscle (resistance to stretch) is maintained after excitation has terminated. There is strong evidence that this is because the cross bridges remain attached and that these detach only in response to relaxing nerve impulses (or serotonin). This implies that the closing muscles of bivalves, once they have shortened and entered into a catch, maintain their length without further expenditure of energy. Maintaining a state of tension without energy is advantageous for animals under anaerobic conditions, especially if the tension is to be maintained for prolonged periods.

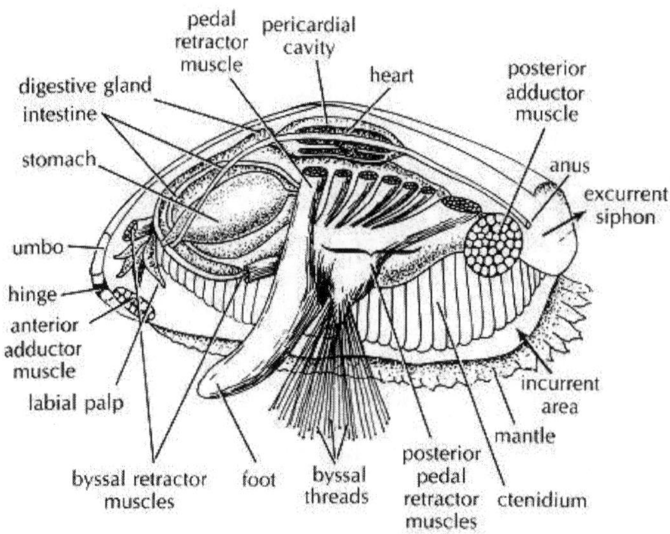

Fig 1.16 - Molluscs muscles

3.8.2. Crustacean muscles

All crustacean muscles are striated, with sarcomeres. Live crabs or lobsters have an impressive force they can exert with their claws. This is not because the strength of the muscle is extraordinarily high but instead because of the anatomical arrangement of the fibres. In the claws of the crab, the muscles found are pinnate. The fibres are arranged at an angle to the direction of the pull instead of being parallel, which significantly increases their mechanical advantage.

In the enclosed space of the claw, a pinnate muscle of a given volume can have more and shorter fibres than a parallel-fibred muscle. The increased force, attributable to this mechanical arrangement, is gained at the expense of the distance over which the attachment can be moved. In the crab claw, the closing muscle can exert, because of the pinnate arrangement, about twice as much force as it could if it were parallel fibred.

Another advantage of the pinnate muscle in the crab claw is related to the confined space in which it works. As a parallel-fibred muscle contracts, its cross section increases, and in the rigid enclosure of the claw, this would cause difficulties. Because of the arrangement of the pinnate muscle, however, the cavity of the crab claw can be filled with muscle, for as the fibres shorten, their angle changes so that there is space for thickening. This is impossible with a parallel-fibred muscle in a limited space.

A pinnate muscle can perform only the same amount of work as a parallel-fibred muscle of equal volume because a corresponding decrease in shortening distance offsets the increased force of contraction. It should be noted that pinnate muscles are not unique to crustaceans; they also occur quite commonly in vertebrates.

The most interesting aspect of the crustacean muscle is the multiple innervations. Two or more nerve fibres may innervate individual muscle fibres. In addition to multiple nerves that stimulate the muscle to contract, many arthropod muscles also have inhibitory nerves whose stimulation causes the muscle to relax if it is already in a state of contraction.

Another crustacean characteristic is that only a few or a single axon often innervates a whole muscle. The crustacean motor axons can be broadly categorized as phasic and tonic based on their general activity patterns and physiological properties. The terms "phasic" and "tonic" are felt to be preferable to the traditional terminology of "fast" and "slow.

The most striking characteristic of crustacean muscle is that two or more different nerve fibres supply it. Often, there is a fast excitatory fibre, a slow excitatory fibre, and an inhibitory nerve fibre. In many muscles, most fibres are innervated by the fast nerve. Combined with the inhibitory fibres, this permits a wide range of contractions.

3.8.3. Comparison of mammalian and crustacean systems

	Mammalian	Crustacean
Muscle fibres	Almost all twitch fibres are fast-twitch red or white, slow-twitch intermediate; Biochemical variation but no sarcomere variation.	Range from short-sarcomere twitch fibres to long-sarcomere tonic fibres, with variable mitochondrial content
Motor axons	Phasic and tonic motoneurons	Phasic and tonic motor axons, inhibitory axons
Innervation	One end plate per muscle fibre (except extraocular muscles)	Multiterminal, usually poly-neuronal
Matching of nerve and muscle	Phasic motoneurons to fast-twitch muscle fibres; Tonic motoneurons to slow-twitch muscle fibres	Phasic motor axons go preferentially to phasic muscle fibres; tonic motor axons to all types, but often preferential innervation of tonic muscle fibres.
Effect of denervation	Nerve terminals degenerate; muscle atrophies; denervation super sensitivity, etc.	Nerve terminals survive; muscle fibres unchanged; no marked structural alterations unless nerve terminals degenerate.
Neural trophic influence on muscle fibre properties	Cross-reinnervation of slow and fast muscles leads to the conversion of biochemical and contractile properties; changes in motoneuron activity can influence muscle fibre properties	No experimental evidence

3.9. Myosepta

Most edible fish products are derived from the skeletal muscles (flesh), representing more than 50 percent of the total body mass of these animals. The skeletal muscles of fish differ from those of mammals and birds in that they are composed mainly of stacks of short bundles of muscle fibres called myomeres. In fishes, the myomeres are W-shaped. The myomeres are separated by thin horizontal (myosepta) and vertical (myocommata). EPAXIALS = above the septum. HYPAXIALS = below the septum. The unique structure and thin connective tissue sheaths of fish muscle give the meat its characteristic soft and flaky texture. Myosepta have been hypothesized to function as transmitters of muscular forces to axial structures during swimming.

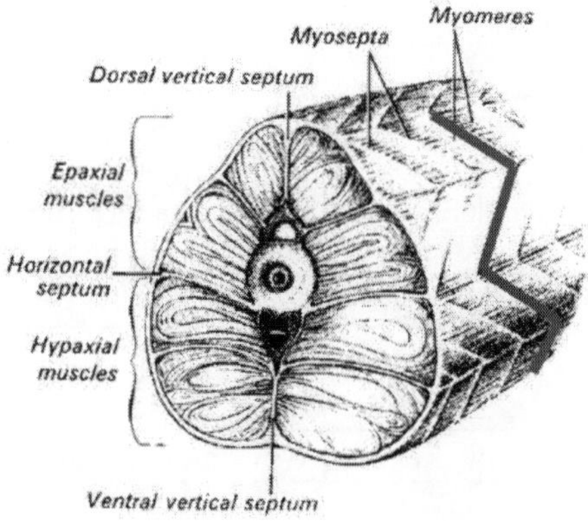

Fig 1.17 - Myosepta

3.10. Dark muscle

Fish muscles consist of light and dark muscle, and the dark muscle is a band of dark tissue that lies beneath the skin throughout the body. In tuna, dark muscle is also located near the backbone. Although the relationship between light and dark muscle generally varies with the fish activity, fatty fish contains a higher proportion of dark muscle, reaching values up to 48%, like in yellowfin tuna. This is because fatty fish are migratory species and need more fat, glycogen and myoglobin for their long journeys. There are many differences in the chemical composition of the two types of fish muscle, but the most important is the high fat and haempigments content in dark muscle. In demersal fish, i.e., species that feed on the bottom and only move periodically, the amount of dark muscle is minimal.

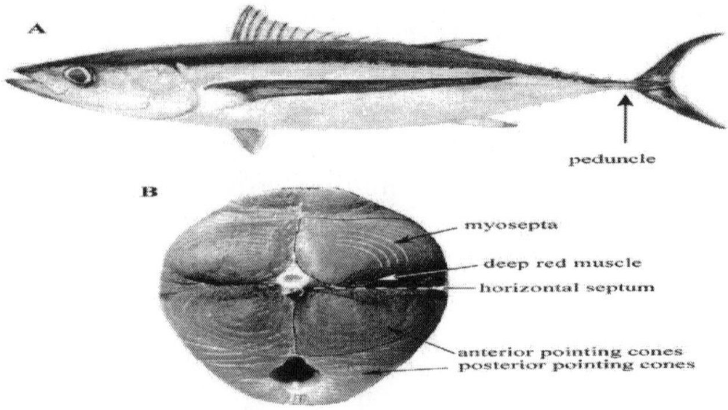

Fig 1.18 – Red muscle

Difference Between White Muscle and Red Muscle Fibres	
Red Muscle Fibres	**White Muscle Fibres**
Definition	
Red muscles are skeletal muscle dense with capillaries and rich in myoglobin and mitochondria.	White muscles are also a type of skeletal muscle, but they contain lower amounts of mitochondria and are myoglobin-poor
Mitochondria	
More in number	Comparatively less in number
Sarcoplasmic Reticulum	
Red muscles have reduced quantities of SR	White muscles have more SR
Fatigue	
Can perform aerobic oxidation without accumulating much lactic acid – hence, can contract for long periods.	Performs anaerobic oxidation and accumulates lactic acid much quicker than red muscles – hence, gets fatigued after short bursts of contraction
Rate of Contractions	
Slow rate of contraction	Fast rate of contraction

3.10.1. Composition and metabolism of dark and white muscle

The energy source for ATP generation in the light muscle is glycogen, whereas the dark muscle may use lipids. A major difference is that the dark muscle contains much more mitochondria than the light muscle, thus enabling the dark muscle to operate an extensive aerobic energy metabolism, resulting in CO_2 and H_2O as the end products. The light muscle, mostly generating energy through anaerobic metabolism, accumulates lactic acid, which has to be transported to the liver for further metabolization. In addition, the dark muscle is reported to possess functions similar to those found in the liver.

The different metabolic patterns found in the two muscle types make the light muscle excellently fitted for intense, short muscle bursts, whereas the dark muscle is designed for continual but weak muscle movements.

In most fish, muscle continues to grow throughout life; therefore, musculature enlargement combines newly formed cells and the increased size of pre-existing fibres. Skeletal muscle includes cephalic, trunk, tail and appendicular musculature. The skeletal muscle of the trunk and tail plays the most important part in locomotion. Appendicular muscles derive from the trunk, and myotomes play a secondary role in locomotion. Appendicular muscles are of two types. 1. Extrinsic - originates on axial skeleton, fascia, or trunk and inserts on girdles or limbs; 2. Intrinsic - originates on girdle or proximal skeletal elements of appendage and insert on more distal elements. Cephalic muscles associated with the opening and closing of the mouth and opercula and the eye and gill movement

3.11. Sound Producing Muscles

A number of fish make humming, grunting or other low-pitched sounds. Such species include toadfish (*Opsanus tau*), squirrel fish (*Holocentrus rufus*) and northern midshipman (*Porichthys notatus*). In these fishes, striated muscles are located symmetrically and completely attached to the wall of the swim bladder. Sound is produced by the rapid muscular contractions of the swim bladder produce the sound in much the same fashion as a stretched taut balloon by dragging a sticky finger across the surface.

3.12. Electric Organs in Fishes

Many fish have special organs that can generate electricity. Electrical organs in fishes are advanced and specialized structures that enable the carrier to generate, store, and release electricity. Only fish have the ability to generate this electricity in the animal kingdom. About 250 species of cartilaginous and bony fish can generate this electricity. This organ has been introduced in fish to meet specific needs. Cartilaginous fish have well-developed electric organs, especially electric rays (*Narcine timlei, N. brunea, Narke diptergia*) and skates. This organ exists in several species of bony fish, such as the *Electrophorus electricus*, which is found in the rivers of South America, and the *Gymnarchus* and *Mormyrus*, a catfish found in the African river.

3.12.1. Origin of Electric Organs in Different Fish:

In skates, *Electrophorus electricus* and Mormyrids (*Gymnarchus* and *Mormyrus*) etc., the tail muscles of fish are transformed to form electric organs. This electric organ is formed by transforming the bronchial muscles adjacent to the gill arch in the *Narke* (= Torpedo), the body muscle in the catfish (*Malapterus*) and the eye muscle in the Stargrazer (*Astroscopus*).

3.12.2. Position of electrical organs in different fish

Different positions of electric organs are observed in different fish. In *Narke*, it is located on both sides of the head along the front edge of the gills and thoracic fins and is connected to the brain through nerves. In electric eels such as *Electrophorus*, *Gymnotus* and *Sternarchus*, it is located laterally from the stem to the end of the tail.

Gnathonemus and *Gymnarchus* are confined to the tail region and are nerves from the spinal cord. The African catfish (Melapterus) is on the skin and muscles along the whole body. In *Astroscopus*, it is located behind the both eyes.

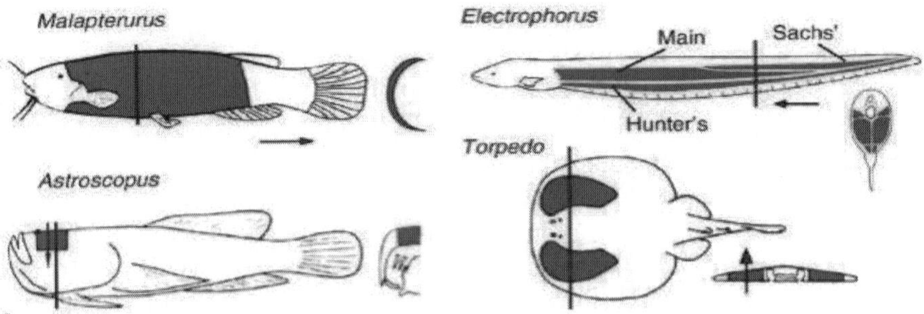

Fig 1.19 - Position of electrical organs in different fish

3.12.3. Function of Electric Organs:

An electric organ is capable of generating electric currents of different voltages. The work of this organ is mainly self-defence, but it also does many other things. Different fish vary in the structure, position and strength of their electric organs due to their various functions. The following are some of the functions of the electric organ:

1. Strong electric organs help to catch prey and protect it from enemy attacks.
2. In *Malapterus* and *Narcine*, this organ has more or less protective function.
3. The use of gentle electric currents has recently been reported to create an electric field around the Mormyrids. If an object tries to break this field, the fish responds to this change immediately. Mormyrids have poorly developed eyes, and many live in muddy water. So, this electric field acts as a warning method. However, it has not been discovered how the fish can detect the breakdown of the electric field. Probably because of their large cerebellum and dorsal fins, special cells are thought to be at work.
4. The electric current helps the fish to determine the direction of movement in the dark. It depends on the change in the electric current around the fish. For this purpose, the lateral line system can detect the flow of electricity.
5. Another function of the electric current is to protect the boundaries of the fish habitat. In the case of some species of Mormyrids, the electric current increases when a fish of the same species comes in contact with another fish.
6. This organ is used to identify the species or sex. These electric signals vary between the two species. If the sensory organs of the fish can detect the pattern of electric current, it helps to identify and locate another member of the same species in the dark.

CHAPTER 4

CIRCULATION

4.1. Introduction

Fishes are cold-blooded aquatic vertebrates found in both saline and fresh water. The circulatory system of fish is responsible for transporting blood, oxygen, hormones and nutrients throughout the body and removing the metabolic waste carried to the excretory organs.

The vascular system in fishes consists of the following components

1. Blood
2. Heart
3. Arterial and venous system

4.2. Blood

Blood is not in direct contact with the tissues or cells. The blood in fish is red and consists of fluid plasma and blood cells or corpuscles. About **2-3%** of the body weight is of blood. Blood is present in the form of watery fluid (90%) or plasma having solutes in the form of proteins like albumin, globulin and fibrinogen, salts, enzymes, antibodies and hormones and consists of blood cells. It also contains the nitrogenous waste products of the body during their transfer to the organs of excretion. Minerals like Ca, Fe, Na, K. Mg, P and Cl are present. The blood clotting factors prothrombin is present. The blood clots rapidly in fish because of the presence of thrombokinase. In the blood of fish, cells like RBC (erythrocytes), WBC (leucocytes) and thrombocytes (blood platelets) are present. RBC (erythrocytes) and WBC are formed from hemocytoblast precursor cells, which may originate from various organs but usually mature after entering the bloodstream. Elasmobranch fishes produce RBC (erythrocytes) from the organ of Leydig (situated in the oesophagus), special tissue around the gonads and especially the spleen. Teleostean hemopoietic (blood-forming) sites are primarily the kidney and spleen. Fish bone has no marrow for hemopoiesis.

RBCs (erythrocytes) are usually the most abundant cells in fish blood (up to 3 million/cubic millimetre). They contain haemoglobin (hb) and largely carry oxygen from the gills to the tissues. Fish RBC is nucleated, contains hb, and shows various sizes among different species. Elasmobranchs have larger red blood cells and fewer in number than teleosts. Mature red blood cells of fishes are oval and tiny and range in long diameter from 7μ in many fishes, such as in certain wrasses, to 36μ in the African lungfish. Enucleated RBC is rare, but the proportion of immature, roundish RBC varies considerably among fish. The more active forms tend to have more erythrocytes than sedentary forms.

Oxygen transport in the blood depends on the iron compound, and hb is the respiratory pigment of the blood. The hb content in the blood varies with the number of erythrocytes present. The number of red blood cells in fish varies depending on the life history stage, habits, and environmental conditions.

Apart from erythrocytes with their respiratory pigment, fish blood contains several types of colourless or white cells, all ovoid to spheroid in shape. The white blood cells vary between 20,000 to 1,50,000 per cubic millimetre of blood in different groups of fish. The WBCs are of 2 types called granulocytes and agranulocytes.

Based on their staining reactions, **granulocytes** are subdivided into neutrophils, acidophils and basophils. Acidophils (eosinophils) and basophils are rare in fish blood. Of these, neutrophils and acidophils are phagocytes.

Agranulocytes are distinguished as lymphocytes and monocytes and are present in large numbers in blood.

Lymphocytes can vary in size (4.5μm to 12 μm in diameter) among species. The number of lymphocytes varies among species and can vary with season. Teleost lymphocytes appear to be produced by both the thymus and kidney. Lymphocytes appear to differentiate into 2 populations, one concerned with the production of antibodies and the other with cellular immunity.

Monocytes comprise a small proportion of the WBC population unless foreign substances are present in the tissues or bloodstream. Monocytes originate in the kidney and are also termed macrophages. These are important in cellular defence, and their role is to eat (phagocytize) anything they come across in the blood that might harm the fish, such as bacteria or parasite larvae.

WBCs function in various ways by riding away foreign material (including invading pathogens) and providing a mechanism for blood clotting. Measurements of change in total WBC number or the percentage of various types can lead to a better understanding of the physiological or pathological state of the animal. Circulating WBC numbers can also vary throughout the year in some fish species. WBC counts generally show an inverse relationship with fish condition or health. Sick individuals presumably make more WBC to synthesize antibodies, phagocytize bacteria, etc. (phagocytic activity – engulfing foreign cells).

The thrombocytes or platelets

These are small, spindle-shaped cells and account for about ½ of all leucocytes in fish. These are involved in blood clotting.

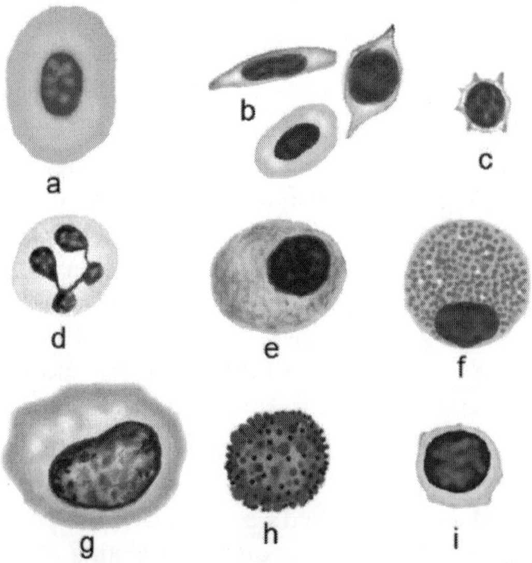

Fig 1.19 - Erythrocyte (a), normal (b) and activated (c) thrombocytes, neutrophil of trout (d), heterophil of shark (e), eosinophil (f), monocyte (g), basophil (h), and lymphocyte (i)

Functions of plasma

- Transport of products of digestion from the small intestine to various tissues.
- Transport of excretory products from tissues to excretory organs
- Transport of hormones from endocrine glands to target organs
- Maintenance of temperature by distribution of heat all over the body
- Provides factors for clotting of blood (fibrinogen)
- Retention of fluids in the body (through plasma proteins)
- Provides body immunity through antibodies (immunoglobulins), which are made by 1 kind of WBC and then released into the plasma

Polymorphism in haemoglobin

Haemoglobin is the red colour respiratory pigment present in the RBC. Hb is present in most fishes but varies depending on the number of red cells in the blood. A few Antarctic ice fishes are without hb in their blood. These species can survive and live without hb because they are extremely sluggish in habit and have low oxygen requirements. They possess a large heart and blood volume as special adaptive features.

1. Monomeric found in Agnatha and hag fishes consisting of a single polypeptide molecule.

   (The polypeptide molecule is the spiral structure (alpha-helix) that forms most of the polypeptide molecules, the building units of the proteins).

2. Tetrameric hb has 4 chains of amino acids (2 alpha and 2 beta chains) found in higher fishes.

The tetrameric hb has a molecular weight of 65,000 daltons and is of several kinds. More than 1 type may be found in the same species. The blood of rainbow trout consists of 4 kinds of hb; goldfish has 3 kinds, and eels have only 2 kinds of hb. Each kind of hb performs different functions, and several combinations have evolved in different species to suit the environmental conditions in which they live. In the migratory eel, 2 kinds of hbs are present. Of these, 1 kind has a high oxygen affinity in seawater, while the other has a high oxygen affinity in freshwater. Thus, the eel is adapted to live in water of different salinity and can maintain a nearly constant level of blood oxygen in fresh and saline water.

It has been found that the goldfish acclimatized to 2°C has 2 kinds of hb, while 3 types are present in those living at 20-35 °C. It has also been shown that this third type of hb appears and disappears with a change in temperature. The 3rd type of hb is formed by rearranging the alpha and beta chains of the other 2 types and is not synthesized fresh. Polymorphism in hb has been associated with the level of activity of the fish. Several factors, such as pH, carbon dioxide concentration and temperature, influence the blood oxygen affinity. Thus, polymorphism in hb appears to be an adaptation to maintain the required blood oxygen level under various conditions.

4.3. Circulatory system in fish

Fish have a closed circulatory system, i.e., blood travels across the body through the network of blood vessels. Fish heart carry only deoxygenated blood; thus, it is called a Venous heart.

There are 3 basic types of heart found in animals: 2 chambered heart, 3 chambered heart, and 4 chambered heart. Fish have 2 chambers, one atrium and one ventricle. Amphibians and reptiles have 3 chambers: 2 atria and a ventricle. Crocodiles are the one reptilian exception, as they have 4 chambers (2 atria, 2 ventricles). Birds and mammals have 4 chambers (2 atria and 2 ventricles)

Two types of circulatory systems are found in animals: The closed circulatory system and the open circulatory system. Vertebrates and a few invertebrates have a closed circulatory system. Closed circulatory systems have the blood closed at all times within vessels of different sizes and wall thicknesses. In this system, blood is pumped by the heart through vessels. The open circulatory system is common to molluscs and arthropods. In the open circulatory systems (evolved in crustaceans, insects, molluscs and other invertebrates), the heart pumps blood into a haemocoel, with the blood diffusing back to the circulatory system between cells. Blood is pumped by the heart into the body cavities, where the blood surrounds tissues.

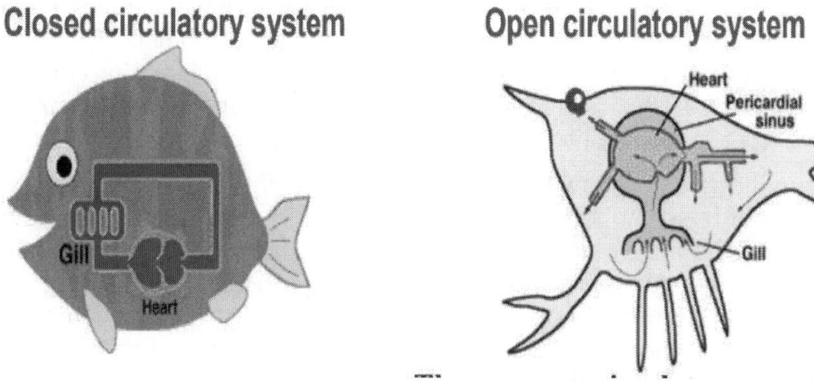

Fig 1.20 - Closed circulatory system and Open circulatory system

Sl.No.	Closed systems	Open systems
1.	Usually, high-pressure systems	Usually, low-pressure systems
2.	Blood is conveyed directly to organs.	Similar to closed systems
3.	Distribution to different organs can be regulated	Distribution of blood not ready regulated
4.	Blood returns to the heart rapidly.	Blood returns to the heart slowly.

Fish have a **closed circulatory system** with a heart that pumps blood around the body in a single loop-from the heart to the gills, from the gills to the rest of the body, and then back to the heart. The fish's heart consists of four parts: the sinus venous, atrium, ventricle, and the bulbous arteriosus. The circulatory system of fish is quite simple and consists of Heart, Blood and Blood Vessels.

The heart is a simple muscular structure located below the pharynx and immediately behind the gills. The pericardial membrane or pericardium encloses it. In most fishes, the heart consists of an **atrium, a ventricle,** a sac-like thin-walled structure known as **sinus venosus** and a tube known as **bulbus arteriosus**. Despite containing four parts, the heart of a fish is considered two-chambered.

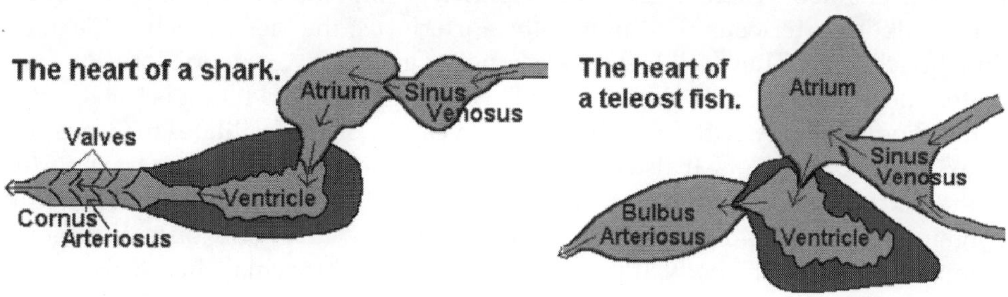

Fig 1.21 – Heart of shark and teleost fish

Sinus venosus: The first chamber is called the sinus venosus, the preliminary collecting chamber. It receives blood through ducti cuvieri, a pair of hepatic veins, a posterior cardinal and an inferior jugular vein. The sinus opens into the atrium through a sinoatrial aperture guarded by a pair of membranous semi-lunar valves. Each valve has a longer limb extending anteriorly into the auricle.

Atrium: From the sinus venosus the blood flows into the atrium. The atrium is the largest of the chambers and is weakly muscular. The narrow lumen of the atrium is called the atrioventricular canal and continues into the ventricle through a circular atrioventricular aperture guarded by 2 pairs of semi-lunar valves of nearly equal size. Each valve has a shorter limb cemented to the atrioventricular ring, and a longer limb is attached to the wall of the ventricle. The spongy part of the atrium consists of parallel strands of cardiac muscles.

Ventricle: The ventricle has a thick muscular chamber nearly as large as the atrium. It is the workhorse of the heart, and its contractions drive the blood around the body. It leads into the bulbous arteriosus. The opening between the 2 is called the ventricular-bulbus aperture and is guarded by a pair of semi-lunar valves. Each of these valves has a shorter limb attached to the wall of the ventricle and a longer limb attached to the wall of the bulbous. The long limbs of the 2 valves cross each other and become attached to the wall of the opposite side, and the 2 valves hang into the cavity of the ventricle.

Bulbus arteriosus: The last chamber of the fish heart is called the bulbus arteriosus in the teleost, but the conus arteriosus in the elasmobranchs. The difference between these chambers is that the conus arteriosus of sharks and rays contains many valves, while the bulbous arteriosus of bony fish contains none. Both are alike in being primarily elastic and work to reduce the pulsed nature of the blood, leaving the ventricle and giving it a more even, constant flow.

4.4. Circulatory mechanism:

The deoxygenated blood collected by the veins accumulates in the sinus venosus, before entering the heart. Blood first enters the atrium of the heart, which is a large chamber. Then, it enters the ventricle, from where it is pumped into the tube, bulbous arteriosus. Through bulbus arteriosus, the blood reaches the aorta and then the gills. They facilitate the exchange of gases, i.e., oxygen absorption from water and carbon dioxide elimination. Then, the oxygenated blood is transported throughout the body with the help of blood vessels. Blood facilitates the transport of oxygen and nutrients. It also collects carbon dioxide, which is again transported to the heart and then to the gills to be removed from the body. The heart of fish thus contains only venous blood, which passes through the heart only once during one complete circuit in the body. It is called a "venus" or "branchial" heart.

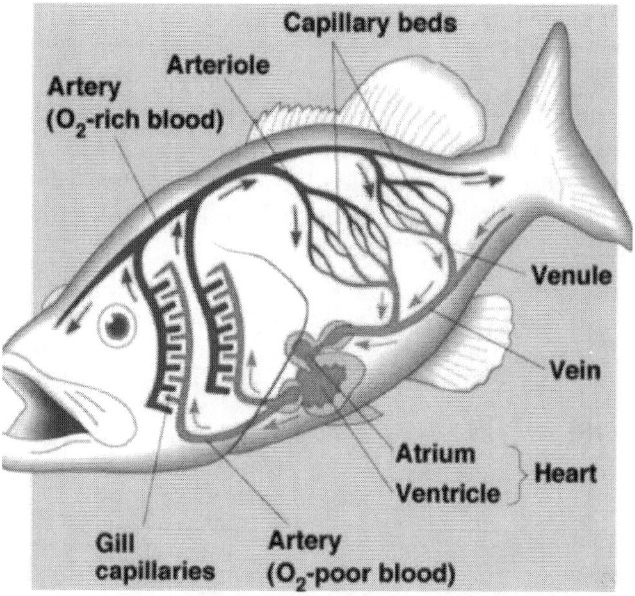

Fig 1.22 - Circulatory mechanism

4.4.1. Blood vessels

The blood vessels are of two types: arteries and veins. The arteries carry oxygenated blood from the heart to the rest of the body, while the veins return deoxygenated blood from the different parts of the body to the heart.

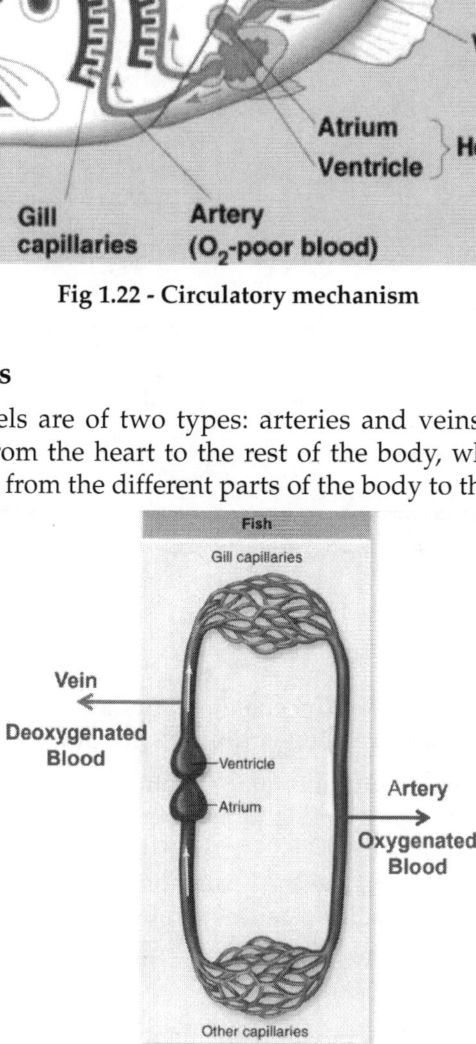

Fig 1.23 – Blood vessels

4.5. Circulatory system in Crustaceans

4.5.1. Circulatory System in Prawn

The prawn circulatory system is open or lacunar, in which the veins and capillaries are absent. The blood vessels open into spaces. These spaces are without a proper epithelial lining and are called lacunae or sinuses. The sinuses together form the body cavity, which looks like a coelom. It is filled with blood known as haemocoel and devoid of epithelial lining.

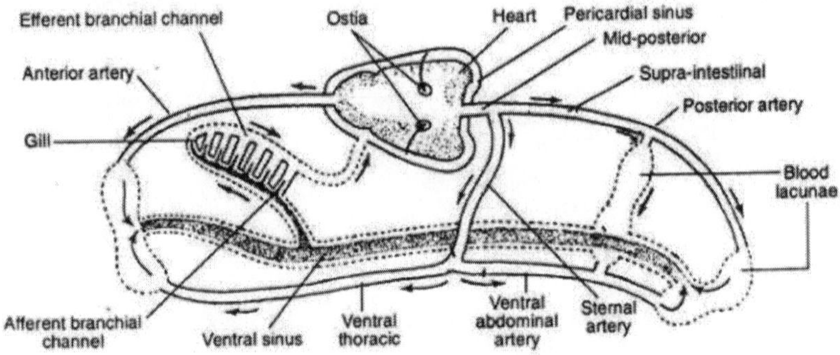

Fig 1.23 - Circulatory system in Prawn

The circulatory system in *Palaemon* consists of

1. Pericardium
2. Heart
3. Arteries
4. Blood Sinuses
5. Blood Channels
6. Blood

Pericardium: It is a thin-walled chamber around the heart that remains filled with haemo-coelomic fluid. The pericardium cavity pumps blood into the heart.

Heart: It is a triangular and muscular structure and is found in the pericardial sinus in the median dorsal part of the thorax. There are 5 pairs of valvular openings, the ostia in the wall of the heart. The 1st pair is in the mid-dorsal region, the 2nd pair in the mid-ventral region, and the 3rd pair on the posterior border. The 4th pair lies anterio-laterally, and 5th pair posterio-laterally.

Arteries: They are thick-walled muscular tubes that arise from the heart and supply blood to different body organs.

Blood Sinuses: Upon reaching the organs, various arteries branch repeatedly into minute branches that do not form capillaries but end in wide space blood sinuses or blood lacunae. All the blood lacunae of the body collectively form a pair of elongated but undefined ventral sinuses. These lie below the hepatopancreas and above the floor of the thorax.

Blood Channels: In Palaemon, there are 6 pairs of afferent branchial channels and 6 pairs of efferent branchial channels. These channels are lacunar in nature. The afferent branchial channels run along the inner surface of the thoracic wall and reach the gills through the gill roots. Here, each channel opens into a transverse connective. 1st branchial channel takes blood to the podobranch and two arthrobranchs. The remaining 5 channels supply the blood to 5 pleurobranches. The blood is oxygenated while passing the gills, and this blood is carried to the pericardium by efferent branchial channels.

Blood: Thin, watery, almost colourless. The colour of blood is white when it is deoxygenated and faintly bluish when it is oxygenated. This is due to the presence of respiratory pigment haemocyanin. There is also a lipochrome pigment called zoonerythin in the blood. Blood has the power of coagulation when it comes in contact with air.

Haemocyanin (Respiratory pigments): It is a high molecular weight protein (75 kD) containing two copper (Cu) atoms. It is dissolved in blood plasma. Deoxygenated blood is white, while oxygenated blood is blue in colour. Normal stability of oxygen binding is achieved only when the unit contains 8 copper atoms (tetramer). Oxygen-carrying capacity is low compared to haemoglobin.

4.5.2. Circulatory system of cephalopods

The *Nautilus* sp. has a heart with four auricles, while the decapoda and octopoda have a heart with only two auricles. This difference is connected with the difference in the number of the ctenidia: four in *Nautilus* (tetrabranchia), two in the *Decapoda* and *Octopoda* (*Dibranchia*). In *Nautilus*, the heart is an almost square sac drawn out to two points on each side; the four auricles that open into the four points of the ventricles are long tubes, more like widened branchial veins than auricles. The solid muscular ventricle of the *Dibranchia* is almost always elongated into a tube. In the *Octopoda*, it lies transversely, the two auricles being in the same plane with the ventricle. In the *Octopsidoe*, the ventricle lies along the longitudinal axis of the body. (i.e.) It is elongated dorsoventrally, and the auricles are at the right angles. The heart of the *Myopsidoe* occupies a position halfway between those just mentioned. The heart described here is the *arterial heart,* which corresponds with the heart of the other Mollusca. It is called arterial to distinguish from the venous hearts, which has been described below.

4.5.3. Circulatory system in Cyclostomes and hagfishes

Cyclostomes - 2-chambered, single circuit, venous and branchial hearts

Hagfishes- The hagfish heart has three chambers in series: the sinus venosus, the atrium, and the ventricle. Some treat a slight microscopic thickening of the base of the ventral aorta as a fourth compartment, the bulbus arteriosus. No major nerves innervate the hagfish heart to stimulate contraction. Instead, filling the sinus venosus by returning blood elicits the Frank-Starling reflex. Accessory 'hearts' (contractile but lack cardiac muscle) like cardinal hearts, caudal hearts, and portal heart (containing cardiac muscles) are blood pumps in different circulation parts.

4.5.4. Accessory hearts

Several fishes have developed accessory pumps that help with venous return. The hagfish has a portal heart that pumps blood from the gut (intestine) through the liver. Both the hagfish and eel possess valved caudal hearts. There are several other types of accessory hearts (valved chambers or vessels) that rely on the contraction of the ventilatory musculature (e.g., the cardinal hearts of hagfish and the branchial hearts found in many fish species) or tail musculature (e.g., the caudal hearts of sharks and eels) to enhance venous return. However, these three types differ from the former in that the musculature that compresses them performs other functions (e.g., in ventilation and locomotion), and their role in venous return is ancillary.

4.5.5. Cardiac output

Cardiac output can be estimated in mammals and birds from Fick's principle. Fick's principle states that the diffusion rate is proportional to the difference in concentration. Similarly, the volume of oxygen consumed per unit time is proportional to the difference in oxygen content between arterial and venous blood. The degree of proportionality depends on the volume of blood pumped per unit time or cardiac output (Q). Therefore, cardiac output (Q) can be calculated from the equation

$$VO_2 = Q\,(C_a - C_v)$$

Where,

VO_2	volume of oxygen consumed (mL/min)
Q	cardiac output (mL/min)
C_a	oxygen content in arterial blood (mL O_2/mL blood)
C_v	oxygen content in venous blood (mL O_2/mL blood)

Cardiac output is the volume of blood pumped by the heart per unit of time and depends on the heartbeat frequency and the volume of blood ejected in one contraction. We can, therefore, write the following equation:

$$Q = f \times S$$

where f is the heartbeat frequency, and S is the volume of blood ejected in one contraction or the stroke volume.

4.5.6. Cardiac Vascular Control

Fish control their hearts and cardiovascular systems with several aneural and neural mechanisms. Venous return is one of the controls that various factors can produce.

4.5.6.1. Aneural Cardiac Vascular Control

This is affected by changes in blood volume, by direct responses of heart muscle to temperature changes and by secretions of various organs. Along with swimming movements in certain species, changes in blood volume will also affect venous return. A rise in temperature causes an increase in the rate of heartbeat in some species. Thus, blood flow increases, and the body receives more oxygen, necessary to maintain a higher metabolic rate in warm water. Certain hormones like catecholamines such as epinephrine also affect aneural control by increasing heart rate.

A decrease in heart rate is also affected by external stimuli, such as the sudden flash of light, touch, mechanical vibration or movement of enemy or prey.

4.5.6.2. Neural Cardiac Vascular Control

The heart in fish is supplied by the cardiac branch of the vagus (x cranial nerve) and is under neural control. It has been shown that vagus nerve stimulation slows the heartbeat in fish.

4.5.7. Myocardial Electrical Activity

Typically, for a vertebrate, most fish hearts are myogenic (i.e., no nervous signal from the brain is necessary for each heartbeat) and show a complex electromyogenic waveform. Although the actual sites of the primary pacemaker nodes are still obscure, evidence gathered for some teleosts suggests islets of pacemaker cells in the sinus venosus and atrium.

A fish electrocardiogram (ECG) is a graphical representation of the electrical activity of the heart during a cardiac cycle. It is obtained by implementing bipolar electrodes under the skin, commonly spanning the pericardial area. The electrodes are wired to an appropriate amplification and display system consisting of an oscilloscope or a physiological recorder. Each peak in the ECG is identified with a letter from P to T corresponding to a specific electrical activity of the heart. The chronological sequence depolarization (contractions) shows through the synchronous movement of blood through the heart. The P-wave represents the electrical excitation (or depolarisation) of the atria, which leads to the contraction of both the atria. The QRS complex represents the depolarisation of the ventricles, which initiates the ventricular contraction. The contraction starts shortly after Q and marks the beginning of the systole. The T-wave represents the return of the ventricles from an excited to a normal state (repolarisation). The end of the T-wave marks the end of systole. Obviously, by counting the number of QRS complexes that occur in a given time period, one can determine the heartbeat rate of an individual.

Wave	Chamber Myocardium Depolarized
V	Sinus Venosus
P	Atrium
QRS	Ventricle
B	Conus ateriosus (elasmobranchs only)

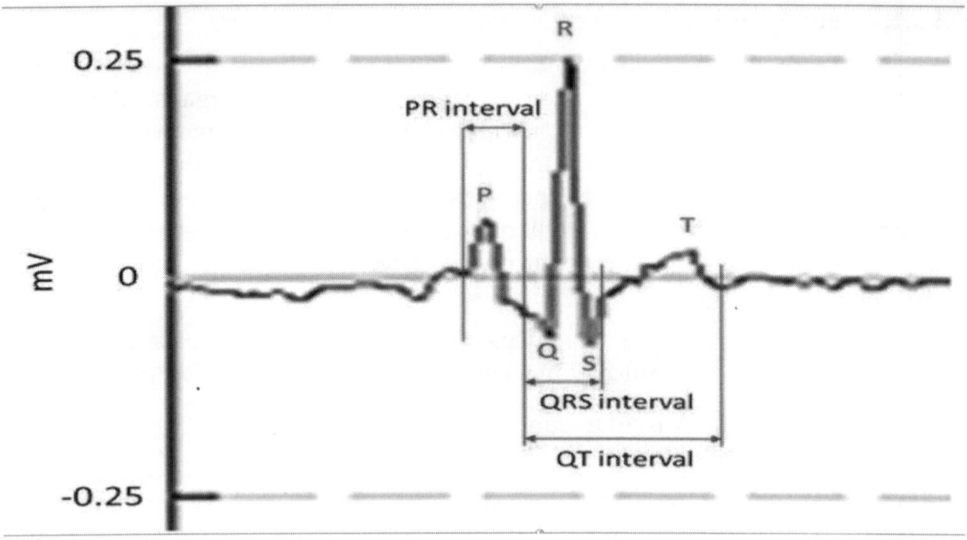

Fig 1.24 - Fish electrocardiogram

UNIT II

CHAPTER 1

EXCRETION

1.1. Introduction

The excretory organs function in waste elimination, osmotic regulation, ionic regulation and the retention of useful organic molecules. The kidney is the primary excretory organ in fishes, as in other vertebrates. In fishes, some excretion also occurs in the gills, where the waste materials produced during metabolism are given off. Skin is also partly excretory in function, serving to remove carbon dioxide. Excretory organs such as kidneys, supplemented by other organs and tissues, maintain the constant internal environment. In other words, they regulate the chemical composition of body fluids by removing metabolic wastes and retaining the proper amounts of water, salts, and nutrients. They accomplish this by a process of selective filtering and processing of the body fluids. Compared with land vertebrates, fishes have a unique problem maintaining their internal environment at a constant concentration of water and dissolved substances, such as salts. A proper balance of water and salt between the body fluids (internal environment) and the surrounding water (homeostasis) of a fish is maintained mainly by the excretory system, especially the kidney.

The vertebrate kidney is divided into pronephros, mesonephros, and metanephros according to the condition of nephros in mature animals. Although pronephros is found in all vertebrates in the embryonic state, only in some animals such as Bellostoma, Myxine, etc., it is an effective kidney in adulthood. However, such kidneys are also found in reptile embryos, birds, and mammals. Fish have mesonephric-type kidneys.

Aquatic animals have no problem removing ammonia. This is why they are called ammonotelic animals. Ammonotelic organisms include crustaceans, polychaets, teleosts, amphibian tadpole larvae, etc.

Some animals live in the water and on land for some time. They convert toxic ammonia in the liver into less toxic ammonia and later into urea. Urea stays in the body longer than ammonia. Animals whose primary excretory substance is urea are called ureotelic animals. Animals such as elasmobranchs, amphibians and mammals are ureotelic animals.

1.2. Excretory System Functions

Constitutes of 4 major processes

1. **Filtration:** pressure-filtering of body fluids producing a filtrate (water, salts, sugars, amino acids, N-wastes)
2. **Reabsorption:** reclaiming valuable solutes (glucose, salts, amino acids) from the filtrate
3. **Secretion**: addition of larger molecules like toxins and other excess solutes from the body fluids to the filtrate
4. **Excretion**: the filtrate leaves the system

1.3. Structure of the kidney

A pair of kidneys lies above the alimentary canal close to the vertebral column. In teleosts, the kidneys are elongated structures usually divided into 2 parts: the head kidney and the trunk kidney. However, in several species, these parts are not demarcated externally.

1.3.1. Kidneys of marine teleosts can be divided into 5 types as described below.

Type I: 2 kidneys are entirely fused, and there is no distinction between the head and trunk parts of the kidney. This is seen in the members of the family Clupeidae.

Type II: The head and trunk regions of the kidneys are clearly demarcated as they are fused in the middle and posterior regions only. E.g. Marine catfishes and eel.

Type III: The kidneys are fused in the posterior region only and the head and trunk potions are clearly demarcated. This is seen in most marine fishes belonging to Mugilidae, Belonidae, Scombridae, Carangidae and Pleuronectidae.

Type IV: Only the extreme posterior parts of the kidneys are fused. Head and trunk regions cannot be distinguished as in the sea horse and pipefish.

Type V: There is no fusion and the kidneys are entirely separate as in Family Lophidae, *Lophius* sp (Goosefish).

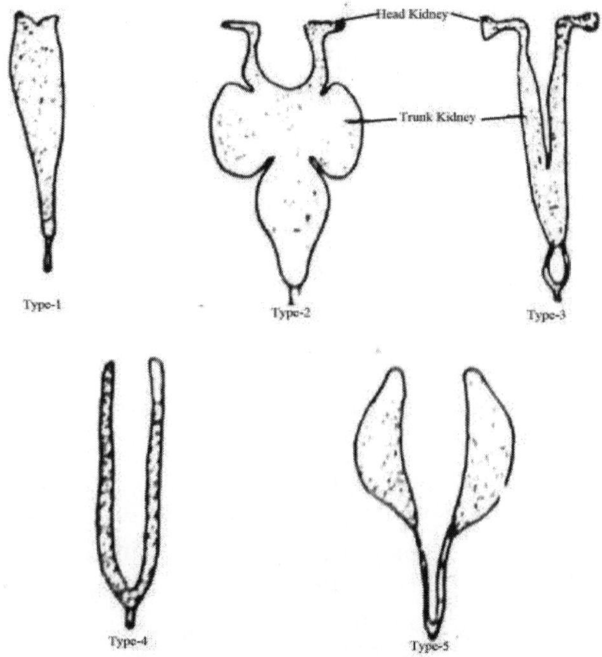

Fig. Different types of Teleostean Kidney (After Ogawa

Fig 2.1 – Types of Teleostean kidney

1.3.2. Kidneys of freshwater teleosts

The first three types of kidneys exist in freshwater teleosts.

Type I: The fully integrated 1st type of kidney is found in Salmon and Trout of the family Salmonidae.

Type II: The second type of kidney is found in fish (carp) of the family Cyprinidae. In this case, the middle and posterior parts of the kidney are integrated. The free segment of the apex forms the head kidney and the integral part forms the trunk kidney. In many species, the middle part of the trunk kidney is broad and the posterior part is gradually narrower. Such kidneys are found in *Cirrhinus, Labeo* and *Puntius.*

Type III: The third type of kidney is found in many species under *Cyprinodontidae, Gasterostidae* and *Cottidae* families. In this case, only the back part is united. Some fish, such as *Mystus, Amiurus* and *Dactylopterus,* have such kidneys. In this case, the head kidney is completely separated from the rest of the kidney.

The kidney usually has two mesonephric ducts, and each crosses along the outer border of the corresponding kidney. It is most clearly seen in the middle and posterior regions of the kidney. The two ducts always meet to form a common duct. Such connections occur at the posterior end of the kidney (*Mystus*) or at a few points between the kidney and the urinary papilla. The two ducts remain separate before entering the urinary bladder (*Labeo, Cirhhinus*).

The common mesonephric duct may enlarge to form the urinary bladder, or it may have a specific sac-like structure on one side of the common mesonephric duct (*Puntius*). In male fish, the urinary bladder is usually exposed through a normal urinogenital opening; in females, it is exposed through a separate urethral opening. The urinogenital opening of the male fish is located at the tip of a papilla.

Urethral openings and genital pores usually exist on the outside, but in some species, a small cutaneous cloaca is formed and exposed in the ducts. Some species have a pair of abdominal pores on either side of the genital pore. The abdominal pore is a special type of opening that is exposed outside the coelom, and the significance of these openings is unknown.

The primary function of the kidney tubule is to remove excess water, salts, waste material and foreign substances from the blood. Blood is filtered under high pressure through the glomerulus and the Bowman's capsule, which work as ultrafilters. The glomerulus comprises a leaky capillary endothelium, a fine porous basement membrane and the epithelium of the Bowman's capsule. These 3 layers form a fine sieve filter. During ultrafiltration under arterial blood pressure, inorganic ions, organic molecules, glucose, urea, and amino acids are filtered into the tubule. The filtrate flows through the tubule due to the filtration pressure and the movements of the cilia of the neck segment and other segments. During this, glucose and salts (Na^+, cl^-) are reabsorbed from the filtrate. Water is also absorbed so that the urine becomes reduced in volume.

1.4. Mechanisms of excretion/ function

The kidneys of freshwater fishes are often larger in relation to body weight than marine fishes. In both groups, the kidney excretes wastes from the body, but freshwater fishes excrete large amounts of water, counteracting the water absorbed through the skin. The kidneys in freshwater fishes also serve as an excretory organ for nitrogenous compounds such as creatine and uric acid. For this, the kidney has many well-developed glomeruli and removes a large quantity of water from the body. Thus, urine is produced around 5-12% of body weight/day in freshwater fishes. The urine is dilute, 20 milli osmoles/L and a urinary bladder is often present for storage.

In marine teleosts, the kidney removes magnesium and sulphate ions along with Na, Cl, K and Ca. Trimethyl amine oxide (TMAO) is also excreted, but there are only traces of ammonia and urea. Very little urine (3 ml/kg/day) is produced, as water has to be conserved to avoid dehydration.

The gills perform the function of the main organs of excretion, removing ammonia and urea along with carbon dioxide. Marine fishes excrete common marine salts via the salt-excreting glands located on the lamellae of the gills, while freshwater fishes excrete salt through the kidneys.

Freshwater fishes tend to lose salt to the environment and must replace it. They get some salt from their food, but the gills and skin inside the mouth absorb salt from water passing through the mouth. This absorption is performed by special cells capable (like those of the kidney) of moving salts against the diffusion gradient. Freshwater fishes drink very little water and take in little water in their food.

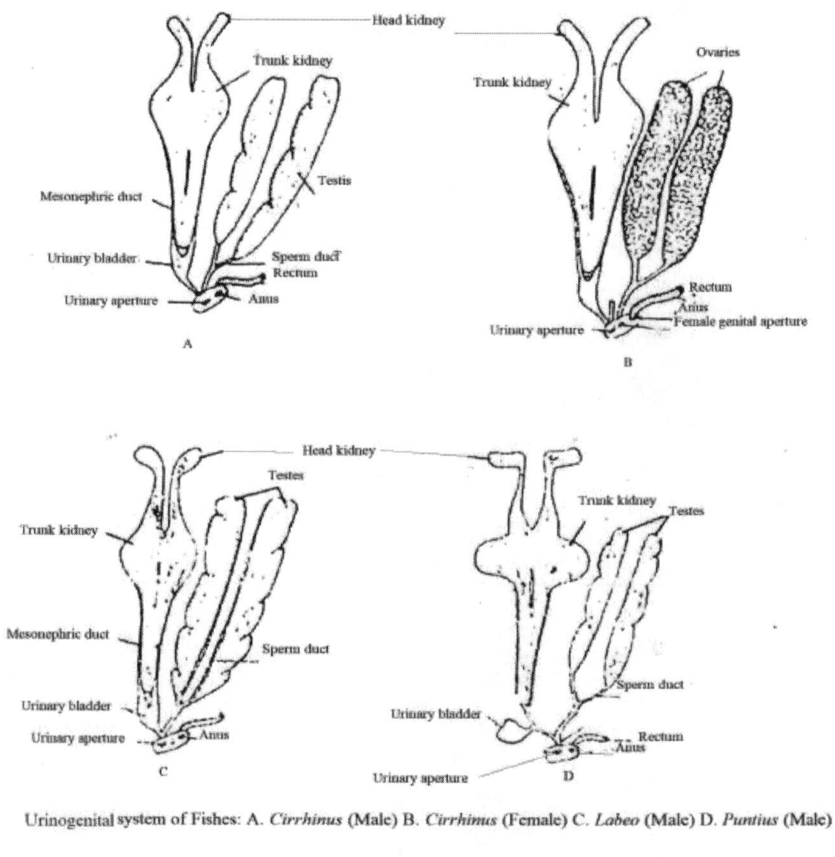

Fig 2.2 – Urinogenital system of fishes A. *Cirrhinus* (Male) B. *Cirrhinus* (Female) C. *Labeo* (Male) D. *Puntius* (Male)

1.5. Histology of Kidney

Histologically, the trunk kidney consists of numerous nephrons. Each nephron consists of a renal corpuscle or malpighian body and ducts, and their inter-ductal space is filled with unevenly expanded lymphoid tissue. The head kidney comprises lymphoid, hematopoietic, intra-renal and chrommafin tissues (suprarenal) and does not contain renal corpuscles and ducts.

In some species, the lymphoid tissue of the head kidney has a small number of nephrons and collecting ducts. However, glomeruli and renal corpuscles are not usually present. The **head kidney has no excretory role**. It originates from the pronephric hematopoietic tissue, but in some teleosts, the mesonephros forms the frontal growth of the head kidney. Renal material is found in the head kidneys in some teleosts. These look like of masonephric type.

In the marine teleosts, the distal segment is probably absent. The renal corpuscles are reduced in number and size in the marine teleosts. The kidneys of marine fishes generally contain fewer, poorly developed vascularized glomeruli than the freshwater teleosts, hence called a **aglomerular** nephron.

Parts of a nephron

A) Renal or malpighian corpuscle

- Glomerulus
- Bowman's capsule

B) Renal Tubules

- Neck region
- 1st proximal segment
- 2nd proximal segment
- Distal segment

C) Collecting duct or tubules

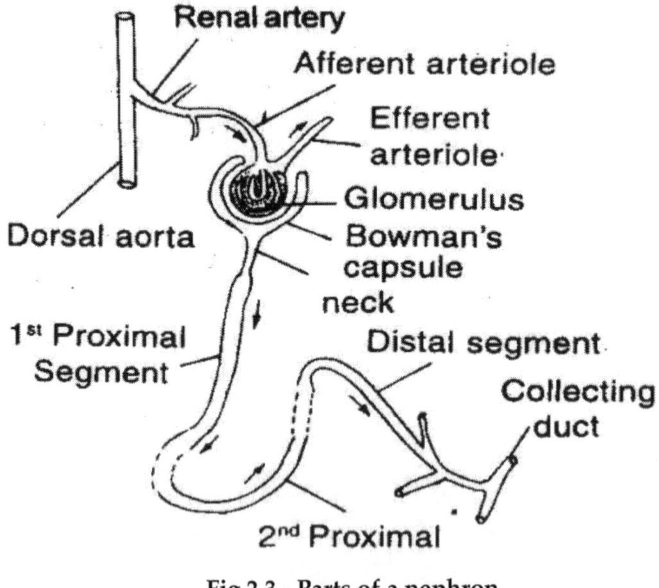

Fig 2.3 - Parts of a nephron

1.6. Excretion in Shellfishes

In many higher crustaceans (decapods), the excretory glands are located in the head. They are called **antennal glands or maxillary glands**, depending on whether they open at the base of the antennae or at the maxillae. The two glands are probably never functional together in the same animal. One may replace the other during development. In the Decapoda, where the antennal gland is well-developed in the adult, the maxillary gland sometimes precedes it in the larva. If the tubule adjacent to the excretory pore is green, the gland is called a **green gland** or **nephridia**. Crustaceans excrete through **nephridia** or **green glands**. Nephridia are invertebrate organs that occur in pairs and function similarly to kidneys. They remove metabolic wastes from an animal's body. They serve 2 functions: osmoregulatory and excretory. They are present in many different invertebrate lines. There are two basic types: metanephridia and protonephridia. A **metanephridium** is a type of excretory gland or nephridium found in many invertebrates such as Annelids, Arthropods and Molluscs. It typically consists of a ciliated funnel opening into the body cavity or coelom connected to a duct, which may be variously glandularized, folded or expanded (vesiculate) and typically opens to the organism's exterior. These ciliated tubules pump water carrying surplus ions, metabolic waste, toxins from food, and useless hormones out of the organism through openings known as nephrostomes. This waste is passed out of the body at the nephridiopore. The primary urine produced by blood filtration (or a similar functioning fluid) is modified into secondary urine through selective reabsorption by the cells lining the metanephridium.

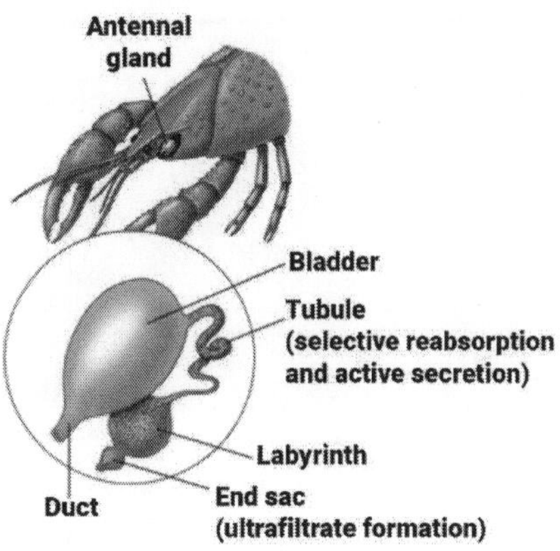

Fig 2.4 - Excretion in Crustaceans

Most insects have an excretory system called **Malpigian tubes,** which collect liquid wastes in the gut. Wastes are crystallized and then excreted to reduce weight. Malacostracans excrete waste fluids mainly through the ducts of the nephridial glands, which are present in the body segments of the second antennae and the maxillae. The ducts open on the basal segments of those head appendages.

There are two different types of arthropod excretory systems. In aquatic arthropods, the end-product of biochemical reactions that metabolise nitrogen is ammonia, which is so toxic that it needs to be diluted as much as possible with water. The ammonia is then eliminated via any permeable membrane, mainly through the gills. All crustaceans use this system, and its high-water consumption may be responsible for the relative lack of success of crustaceans as land animals. Various groups of terrestrial arthropods have independently developed a different system: the end-product of nitrogen metabolism is uric acid, which can be excreted as dry material; Malpighian tubules filter the uric acid and other nitrogenous waste out of the blood in the hemocoel and dump these materials into the hindgut, from which they are expelled as faeces. Most aquatic arthropods and some terrestrial ones also have organs called nephridia ("little kidneys"), which extract other wastes for excretion, such as urine.

CHAPTER 2

OSMOREGULATION

2.1. Osmoregulation-Introduction

Osmosis is the process by which the solvent enters the area from low density to high density. In this case, two different solution densities are separated by a differential permeable membrane. The kidneys play an important role in the osmoregulation of fish. However, most of the osmoregulatory functions are performed by other organs, such as the gills, skin and intestines.

Osmoregulation is a type of homeostasis that controls the concentration of water and ions. Osmoregulation refers to the ability of a fish to maintain a suitable environment inside the body due to osmotic pressure. As a result, there is always a moderate difference in the concentration of intracellular and extracellular ions.

Numerous techniques have been developed in the body of fish to control the various problems caused by osmosis. The most common strategies are (1) Osmoregulation between intracellular and extracellular cells and (2) Osmoregulation between the extracellular chamber and external environment.

Osmotic exchanges between fish and their environment are of two types, viz.

1. **Obligatory exchanges**-it occur by responding to a physical factor over which the animal has little or no physiological control;
2. **Regulatory exchanges** – Such exchanges are physiologically well-regulated and control internal homeostasis.

2.2. Osmoregulators and Osmoconfirmers

Osmoregulators are animals capable of controlling different intra-osmolarity from the environment. Fish that are actual osmoregulators can control the concentration of body fluids. All these fish migrate to freshwater and saltwater and can control the change in osmotic pressure caused by environmental changes through endocrine techniques.

Osmoconformers are those animals that cannot control the osmotic state of their body fluids but ensure the osmolarity of the surrounding medium. Most fish live in either freshwater or marine. In various physiological processes, waste products are excreted out of the body due to metabolism through the intestines, skin and kidneys in vertebrates. However, the gills of fish and aquatic animals are permeable to water and salt.

Most invertebrates among marine animals are osmoconformers, whereas most vertebrates are osmoregulators.

2.3. Osmoregulation in Freshwater Fishes

The body fluids of freshwater fish are more osmotic than the surrounding aquatic medium (hyperosmotic). Due to this, fish face two types of osmoregulatory problems in the water, such as:

1. Water enters the body due to high osmotic body fluids and differences in osmotic pressure. As a result, the body of the fish swells.
2. Due to the low salt concentration in the surroundings, their bodies constantly lose salt, so freshwater fish have to prevent excess water from entering their bodies and losing excess salt. The kidneys produce large amounts of less-density urine to eliminate excess water.

To counter the continuous inflow of water through gills, freshwater fishes produce a large amount of dilute urine. The kidney has a vast number of well-developed glomeruli (often more than 10,000), which are well vascularized and continuously filter excess water which is eliminated as hypotonic (dilute) urine. The kidneys of freshwater fishes are often larger in relation to body weight than marine fishes and produce copious urine, which is 5-12% of the body weight/day. Freshwater fishes drink very little water and take in little water in their food as a large amount of water enters the body by osmosis, which is more than necessary for the fish.

Freshwater fishes lose salts, especially chlorides, through passive diffusion through membranes. Some salts are lost through faeces and also in urine. Urine also contains some nitrogenous waste such as creatine, creatinine, amino acids, urea and a little ammonia. The quantity of salts lost in a day varies in different species. In salmon (euryhaline species), loss of salts may be up to 17% of the body chlorides, while in goldfish, *Carassius auratus*, it is 55%. However, the loss of salts is kept at a minimum by active reabsorption in the proximal and distal tubules. Loss of salts is partly compensated through food and absorption of salt ions from the surrounding water. This is accomplished mainly by the chloride cells in the gills and oral membrane (mouth). Salt ions absorbed by these special cells at the base of the gill lamellae are Li, Ca, Br, Cl, SO_4 and PO_4. The entire mechanism of salt balance is under hormonal control.

2.4. Osmoregulation in Marine Fishes

Marine fish experience the opposite effect compared to freshwater fish because their internal salt content is less than that of the surrounding water. Marine teleosts are **hypoosmotic** (body fluids 1/3 of that seawater) and live in a medium-high concentration of salts. The skin of marine fishes, as in freshwater species, is impermeable and osmotic exchanges occur through the gills and oral membrane. In other words, a marine fish has an internal osmotic concentration lower than the surrounding seawater, so it tends to lose water and gain salt. It actively excretes salt out from the gills. Marine fishes face the problem of continuous loss of water and gain of salts. To prevent osmotic dehydration or maintain their water balance, marine fishes drink large quantities of seawater, which further increases the body's salt content. Urine output is considerably reduced to conserve water. The glomeruli are reduced or become non-functional. The distal segment of the tubule is absent. Thus, only a tiny amount of concentrated urine is produced, and most water is reabsorbed in the tubule.

As marine fish drink seawater, Na and Cl ions are actively absorbed through the gut. Many species are reported to drink water >25% of their body weight/day, of which 80% is absorbed. This excess of salt ions is excreted into the water by the special cells in the gills, oral membrane and inner surface of the operculum. Most nitrogenous waste in marine fishes appears to be secreted by the gills as ammonia.

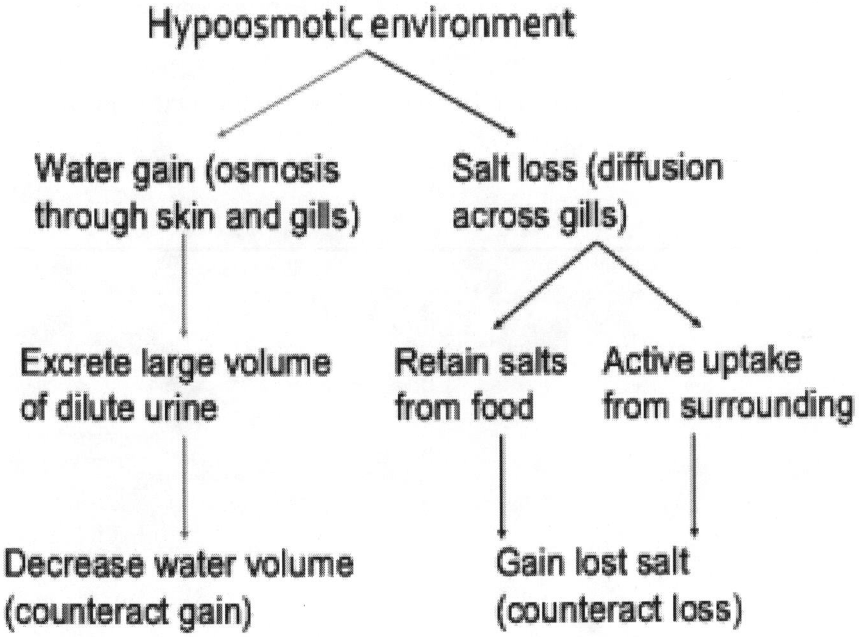

Fig 2.5 - Osmoregulation in Hypoosmotic environment (Freshwater fish)

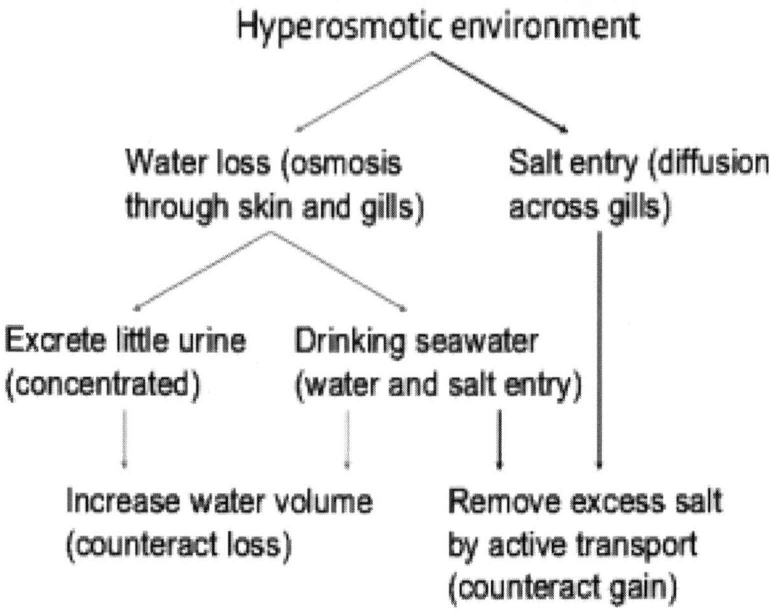

Fig 2.6 - Osmoregulation in Hyperosmotic environment (Marine fish)

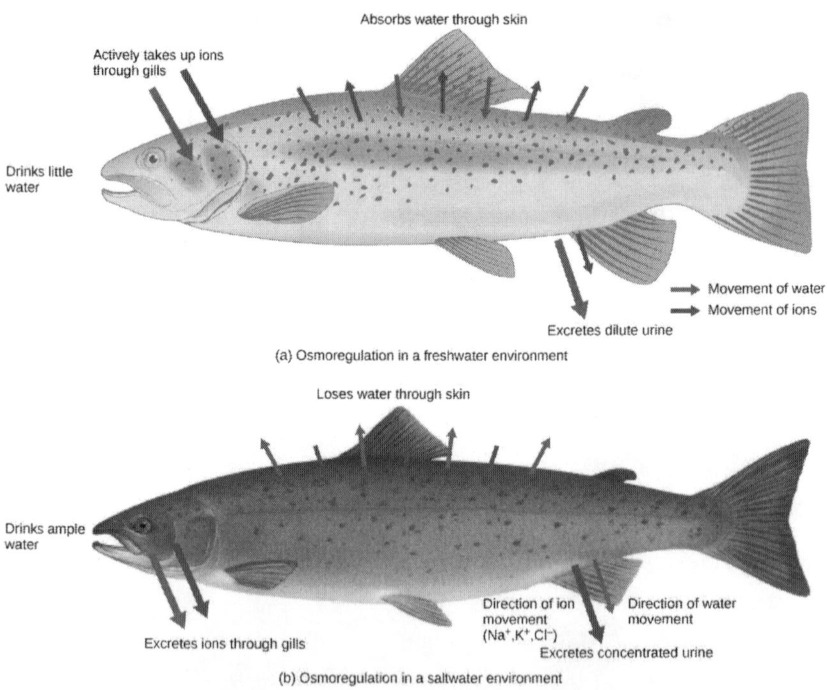

Fig 2.7 – Osmoregulation in freshwater and marine fishes

2.5. Osmoregulatory problem in hagfishes

In hagfishes (marine cyclostomes), the blood is nearly **iso-osmotic** (equal concentrations) to seawater, so there is practically no water movement across the semi-permeable membranes. That is, osmotic work is avoided, as the concentration of Na^+ and Cl^- ions is similar to seawater.

2.6. Osmoregulatory problem in Chondrichthyes or elasmobranchs

The chondrichthyes maintain osmotic blood concentration nearly at the same level as seawater (iso-osmotic). Hence, they are not in danger of dehydration and do not drink seawater. This is achieved by retaining nitrogen-containing compounds, mainly urea and trimethylamine oxide (TMAO), in the blood, which is slightly hyperosmotic to seawater. Consequently, these fishes gain water through osmotic diffusion across the gills and do not drink seawater. The kidney has large glomeruli, as in freshwater fish; filtration forms urine. However, most of the urea in the filtrate is reabsorbed by special segments of the urinary tubule. The gills of chondrichthyes are impermeable to urea, so they are retained in the blood in large quantities to raise the osmotic concentration of blood to a slightly higher level than seawater. It is reported that the average urea content in the blood of chondrichthyes is 1000 – 2000 mg% and that of urine 230 mg%. The osmotic balance in these fishes is maintained by the retention of high urea levels in the blood. Freshwater sharks have a lowered concentration of urea in the blood. Salts are excreted in the faeces and in the urine. Some salts are also reabsorbed in the kidney tubules, but gills do not excrete salts. Gills of chondrichthyes do not have special salt-excreting cells. They achieve salt balance by secreting a fluid containing a higher Na+ and Cl- ions concentration from a special excretory organ, the rectal gland located at the alimentary canal's end.

The rectal gland is a structure found in marine elasmobranchs that helps in osmoregulation by excreting NaCl. It consists of a mass of tubules composed of cells very similar in ultrastructure to chloride cells of fish gills.

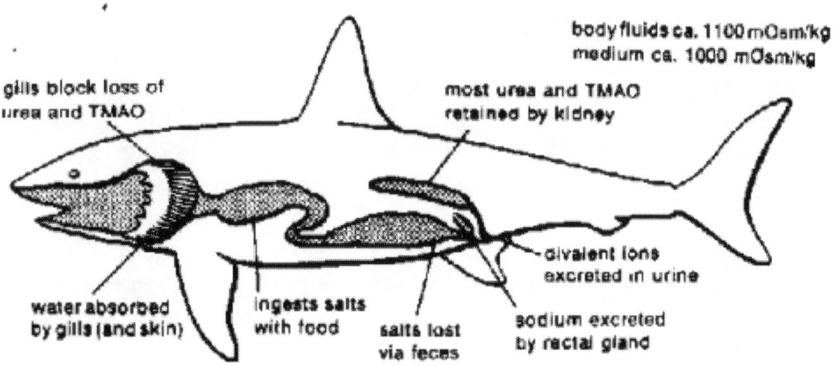

Fig 2.8 - Osmoregulatory in Chondrichthyes

2.7. Osmoregulation in migratory fishes

Migratory fish are hyperosmotic to their environment during some periods and hypoosmotic at other times.

To complete their life cycle, amphihaline fishes show remarkable adaptation in migrating from the sea to freshwater bodies or vice versa. They have to survive in 2 different environments for the sake of their progeny. There are 2 distinct types of amphihaline fish. One lives in seawater, grows there, and after reproductively maturing, starts migrating to breed in riverine or land-locked freshwater. The young born in freshwater migrate into seawater and live there until they are, in turn, mature and ready to spawn in freshwater. These fish, e.g. salmon and hilsa are known as amphihaline potemotocous.

Catadromous fish are another type of amphihaline fish that lives in riverine or lake water when reproductively mature. They migrate into the sea to breed there. These kinds of fish are known as amphihaline thalassotocous. E.g., eel.

In both anadromous and catadromous, there is a multifaceted adaptation to osmoregulation. These fish have glomerular kidneys, which can be adapted by producing different amounts of urine in different salinities. In addition, fish have gills and oral membranes that can absorb and excrete specific ions.

2.7.1. Preparation for life in a changed environment

One of the significant changes is an alteration of osmoregulatory mechanisms, which leads to an increase in the osmotic pressure of blood that suits marine conditions. Since the osmoregulatory process must function actively, the fishes utilize most of their energy for this purpose. Their downstream migration is strictly passive to conserve energy. The preparation of marine life is apparent in the thickening of the layers of skin in eels, which helps them survive in the new marine environment. Also, there is an appreciable increase in the number of gill tissue cells that secrete chlorides. These cells multiply several folds a few days after the fish enters seawater, enabling it to resist salt invasion that threatens it in the marine environment. Silver eel, in its catadromous migration, shows a higher rate of intestinal water flux and a lower rate of sodium penetration through the gill than that observed in the yellow eel in freshwater. The silver eel successfully regulates salinity with its increased aptitude to excrete salts. Water loss is compensated by taking in water. Factors that control the rate of water intake in fish are unknown. Perhaps they feel thirsty. With all these adaptive physiological changes, which are invariably controlled and monitored by the endocrine gland, eels migrate towards the direction of increasing salinity and successfully spawn in the warmest part of the deeper oceans.

A profound modification of the physiological mechanism of osmoregulation is necessary depending on whether the salmon or hilsa live in salt or freshwater. This modification appearing at a precise life cycle stage is one of the determining factors for migration. While approaching coastal waters to begin its upstream migration, an apparent increase in the concentration of salmon's blood occurs, indicating a change in the osmoregulatory mechanism that is generally suitable for the marine environment. The same may be true in the case of hilsa.

The young salmon, born in freshwater, spends several years in these surroundings and undergoes significant morphological, biochemical and physiological modifications before downstream migration to marine feeding grounds. The smolt develops a more pronounced tolerance to salinity than the parr because of the increase in chloride cells in the gills. The activity of 2 enzymes, namely succinic-dehydrogenase and salt-linked ATPase associated with chloride cells, permits the elimination of excess salt through the digestive tract when fish enter the sea. Salmon thus gives hyperactivity to lose chloride and sodium ions in marine water.

2.8. Factors controlling ion-osmoregulation in amphihaline fish

Two hormones are found to be very important for the shift in osmoregulation from seawater to freshwater or vice versa. One is cortisol, a cortico-steroid hormone released from the adrenal cortex. Fish do not have adrenal glands; their corticoid tissues are dispersed in the head kidney region known as the interrenal, which secretes cortisol. Inter-renalectomy (removal of the inter-renal organ) seriously inhibits smoltification. Inter-renalectomised mature eels do not show silver colour, and their transfer to dilute seawater results in death.

On the other hand, administering cortisol allows them to survive in seawater. Cortisol, therefore, is an essential hormone that enables the adaptation of these fishes to marine life. Cortisol also has a role in facilitating adaptation to seawater by increasing the absorption of water in the intestine with concomitant selective exclusion of ions and by stimulating the reabsorption of water from the urinary bladder.

The other important hormone is prolactin, which is required for freshwater life. Prolactin induces the mucous secretion that covers gill filaments and the body surface, inhibiting salts from going out and water from entering. These are the 2 essential requirements for seawater fish as they live in a hypoosmotic surroundings. Prolactin also regulates water and ion movement through the intestine and urinary tract. Prolactin, therefore, is very important in maintaining ion balance in freshwater. Besides cortisol and prolactin, thyroid hormone, growth hormone, and insulin have also been shown to be involved in the ion-osmoregulation of amphihaline fish.

2.9. Osmoregulation in shellfishes

Animals, such as crabs, have an internal salt concentration similar to the surrounding ocean. Such animals are known as osmoconformers, and there is little water transport between the inside of the animal and the isotonic outside environment.

Crustaceans have a pair of antennary or green glands and a median renal sac for osmoregulation and excretion.

CHAPTER 3

SENSE ORGANS

3.1. Introduction

Specialized organs receive physical and chemical stimuli from the environment called, sense or sensory organs. Fishes possess several sense organs for smell, sight, hearing, taste, touch, temperature etc. The sense organs are associated with the nervous system and transmit the environmental stimuli to the central nervous system.

3.2. Sense organs

1. Eye
2. Acoustico-Lateralis System
   - Internal Ear
   - Lateral line system
3. Chemo-sensory organs
   - Gustatory (taste buds)
   - Olfactory (olfactory rosette)
4. Electro-receptor organs
   - Ampullary organ
   - Tuberous organ

The nervous system of fishes is poorly developed compared to that of other vertebrates. A bony fish's brain is divided into the forebrain, the midbrain, and the hindbrain. The forebrain is responsible for the bony fish's ability to smell. Bony fishes that have a perfect sense of smell. Eels have an enlarged forebrain. The midbrain processes vision, learning and motor responses. Blind bony fishes, such as blind cave fishes in Amblyopsidae, have a reduced midbrain. The hindbrain (medulla oblongata and cerebellum) coordinates movement, muscle tone, and balance. Usually, fast-swimming bony fishes have an enlarged hindbrain.

The brain of the fish is divided into several anatomical and functional parts, all closely interconnected, but each serves as the primary centre of integrating particular kinds of responses and activities. Several of these centres or parts are primarily associated with one type of sensory perception, such as sight, hearing, or smell (olfaction).

The sense organs receive physical or chemical stimuli from the environment. Physical changes in heat flow or touch are felt through skin receptors. Visual stimuli involve light intensity and quality changes, and acoustical stimuli are received through the inner ear or lateral line. Chemical stimuli are those experienced through either smell or taste organs.

3.2.1. Eye

The eye is the photoreceptor organ in fishes and resembles other vertebrates in structure. The eye in most fishes is lidless, but some sharks (Galeorhinus) possess a lid-like nictitating membrane. Fishes have monocular vision. The field of view of each eye is considerably enlarged and extended forward and continues laterally to almost behind the fish Structure: i) Cornea, ii) Iris, iii) Lense, iv) Eyeball.

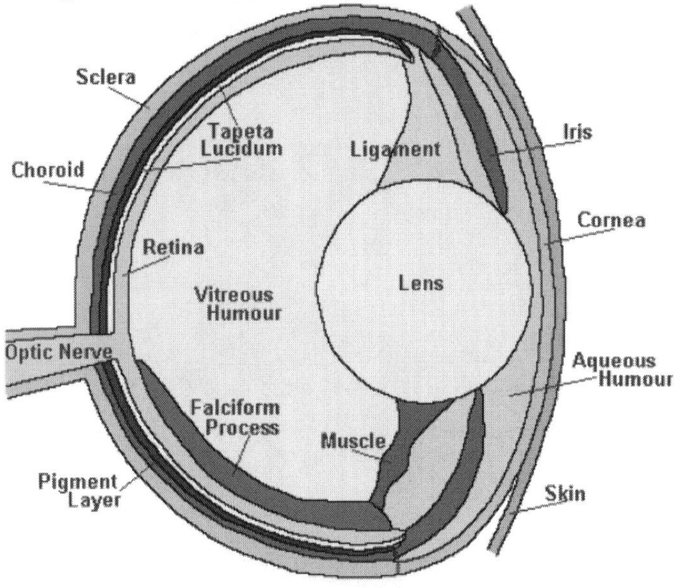

Fig 2.9 – Teleost eye

In general, fishes living in dark and dim water habitats have large eyes, allowing them to absorb as much light as possible in the dark. In some, if they have specialized in some compensatory way so that another sense (such as smell) is dominant, the eyes will often be reduced. Fishes living in brightly lighted shallow waters often will have relatively small but efficient eyes and probably have colour vision. Specific visual cells are specialized to particular wavelengths and intensities. The pupils of some species of bony fishes, such as eels, contract and dilate depending on light conditions. In most bony fishes, however, pupils cannot contract or dilate. Cyclostomes have less elaborate eyes than other fishes, with skin stretched over the eyeball, perhaps making their vision less effective.

Most fishes have a spherical lens and accommodate their vision to far or near subjects by moving the lens within the eyeball. A few sharks accommodate by changing the shape of the lens, as in land vertebrates. Those fishes depend heavily on the eyes and have especially strong muscles for accommodation. Most fishes see well despite the restrictions imposed by frequent turbidity of the water and by light refraction. Experimental evidence indicates that many shallow-water fishes, if not all, have colour vision and see some colours well, but some bottom-dwelling shore fishes live in areas where the water is sufficiently deep to filter out most, if not all colours and these fishes never see colours. When tested in shallow water, they cannot respond to colour differences.

In some species, the eye has a reflective layer called the *tapetum lucidum* behind the retina. The *tapetum lucidum* reflects light through the retina a second time. The mudskipper (family Periophthalmidae) and several other species of bony fishes have excellent eyesight both above and below the surface of the water. The four-eyed fishes (family Anablepidae) swim at the water's surface. Their eyes lie at the water line and are adapted for seeing in air and water. Separate retinae and an asymmetric lens allow these remarkable fish to focus on images above the water and images under water simultaneously.

The eyesight in some species of bony fishes may be well developed. Goldfish (*Carassius auratus*) have excellent visual acuity up to 4.8 m (16 ft.) away. Some species of bony fishes have no eyes. The blind cavefishes (family Amblyopsidae) have no vision perception. Other senses help them find prey. The blind goby (*Typhlogobius californiensis*) has eyes that degenerate as the goby matures.

A notable feature of the typical teleost eye is the cornea of constant thickness. This cornea imposes no optical alterations (convergence or divergence) on incoming light. Thus, all the focus of light occurs at the spherical lens with the highest effective refractive index among the vertebrates. Like other vertebrate eyes, the lens consists of water and structural proteins.

The teleost eye lens protrudes through the pupilar opening in the iris, and the eye bulges from the body's surface. Therefore, the field of view includes a considerable arc forward, continuing laterally to almost directly behind the fish.

Adjustments in focus for near or far vision are accomplished by moving the lens without changing its shape by muscles within the eye. The specific muscles within the eye that are responsible for these movements differ in elasmobranchs and teleosts.

The retina comprises a dense packing of rod and cone cells for the discrimination of light images. Incoming light must penetrate a transparent layer of nerve cells and fibres to reach the photochemically active tips of the rods and cones.

As light strikes the rods or cones, it is absorbed by the light-sensitive pigment (e.g., rhodopsin, porphyropsin) in the cell. These pigments are located in the photoreception area of the retina, which gives it its characteristic purple or pink colouration. In all cases, pigmented light absorption stimulates the retinal cells to send impulses via the optic nerve to the optic lobe on the same side of the brain. The reception and integration of the visual image in the brain would then provide the appropriate motor response (e.g., to the muscles and fins).

The choroid coat underlies the retinal layer and primarily supplies nutrients and oxygen to the retina with its high metabolic demand. Most teleosts possess a pigmented, choroid projection into the posterior portion of the interior of the eye called the falciform process. The falciform process is highly vascularised and probably serves a nutritive function. In addition, a choroid gland is present in most teleosts behind the retina. The arrangement of blood vessels of the choroid makes it eminently suited for oxygen delivery to the retina.

The sclerotic coat or sclera forms a firm layer of the eyeball that is variously reinforced in all groups of fish. Elasmobranchs have cartilaginous elements and calcified plates in this layer for additional reinforcement. The highest of the bony fishes have fibrous, flexible sclerotic coats within the orbit and around the optic nerve, but there may be cartilaginous or even bony supporting elements in the outer portion of the eye that can surround the entire cornea.

Pineal Gland

The pineal gland is dorsally located on the brain and has light sensitivity. Blind cavefish (*Astyanax mexicanus*) can sense light when young, even though their eyes lost their function over a million years of evolution. Scientists have found that the fish larvae can detect an overhead shadow and seek shelter by swimming towards it. The researchers found that the behaviour is driven by the pineal gland, a light-sensitive organ sometimes known as the 'third eye'. This gland contains a light-sensitive pigment called rhodopsin, and when it was removed, the fish's shadow response disappeared.

3.2.2. Acoustico-Lateralis System

The acoustic-lateralis system of fishes senses sounds, vibrations and other water displacements in their environment. It has 2 main components: the inner ear and the neuromast/ lateral line system. Besides sound detection, the fish's inner ear functions to orient or balance the animal, giving it a feeling of the direction in which gravity is acting, even when suspended in lightless pelagic habitats.

3.2.2.1. Internal Ear or Membranous Labyrinth

The nature of sound transmission in water significantly influenced the evolution of hearing in fishes. Because of its greater density, water is a much more efficient conductor of sound pressure waves than air. In water, sound travels more than 4.8 times the speed of sound through air. Sound perception and balance are intimately associated with the senses in a fish. The organs of hearing are entirely internal, located within the skull, on each side of the brain and somewhat behind the eyes. Sound waves, especially those of low frequencies, travel readily through water and impinge directly upon the bones and fluids of the head and body to be transmitted to the hearing organs. A fish's soft body tissue has about the same acoustic density as water. Fishes readily respond to sound; for example, a trout conditioned to escape by the approach of fishermen will take flight upon perceiving footsteps on a stream bank, even if it cannot see the fisherman. However, the range of sound frequencies heard by fishes is greatly restricted compared to humans. It is thought that many fishes crudely communicate with each other by producing sounds in their swim bladders, in their throats by rasping their teeth, and in other ways. The teleosts and elasmobranchs detect sound vibrations through the inner ear and the lateral line.

The ears of a bony fish function to maintain equilibrium, detecting acceleration and hearing. There are no external openings to the ears. Sound waves travel through soft tissue to the ears.

The internal ear is a membranous sac partially divided into an upper chamber utriculus and a lower one called sacculus, having a small outgrowth, the lagena. Three semi-circular canals open into the utriculus. Each canal has a small swelling at the end called the ampulla. The cavities of the utriculus, sacculus and lagena contain a fluid called endolymph and a group of highly sensitive cells (Sensory Hair Cells) in their wall. In the cavity of the ear, solid concretions called otolith are present. Three otoliths are present in three sacs: lapillus, sagitta and asteriscus.

Semi-circular canals and utriculus are mainly responsible for the maintenance of equilibrium, while the sacculus and lagena are sites for the sense of hearing. Among the minnows, catfishes and suckers, the inner ear is connected uniquely to the gas bladder through a chain of bones, the Weberian ossicles.

There is significant variation in hearing sensitivity, bandwidth, and upper-frequency limit among bony fish species. The hearing range of the cod *Gadus morhua* is about 2 to 500 Hz, with peak sensitivity near 20 Hz - probably typical for most bony fish species that lack the adaptations described below.

In some bony fish species, the swim bladder is associated with adaptations for enhanced sound reception at higher frequencies. In some, the swim bladder lies against the ear and acts as an amplifier to enhance sound detection. In other species, such as goldfish (*Carassius auratus*), small bones connect the swim bladder to the ear. The hearing range of the goldfish is about 5 to 2,000 Hz - with peak sensitivity near 400 Hz. Recently, researchers have discovered that the American shad (*Alosa sapidissima*) and certain related species can detect sounds from 200 to 180,000 Hz. The researchers theorized that this ultrasonic hearing is an adaptation for avoiding echolocating dolphins, which typically produce clicks at about 100,000 Hz.

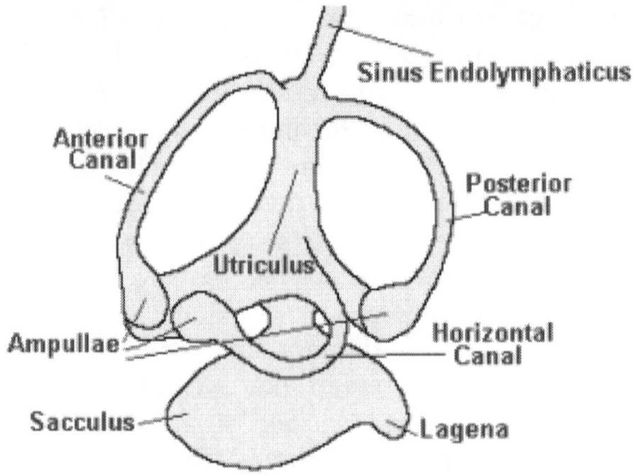

Fig 2.10 - Membranous Labyrinth

3.2.2.2. Lateral line

The lateral line system is an important sensory system in fishes that is absent in other vertebrates (except some amphibians). Like the ear, the lateral line senses vibrations. It functions mainly in detecting low-frequency vibrations and directional water flow and in distance perception. This consists of a series of fluid-filled canals just below the skin of the head and along the sides of a bony fish's body, where it is associated with the scales. The canals are open to the surrounding water through tiny pores. Lateral line canals contain sensory cells (pit organs) that detect changes in pressure. The system allows a fish to sense changes in water currents and pressure, thereby helping it orient itself to the various changes in the physical environment.

The lateral line of fish provides a sense of distant touch. Water movements around the fish can be detected through mechanoreceptors similar to those in the auditory and equilibrium systems. The receptors are called neuromasts. Each neuromast is a group of sensory cells, supporting cells and basal cells. The sensory cells are called hair cells, which possess a bundle of cilia (stereocilia and kinocilium) on the apical surface. The stereocilia are full of microfilaments and become longer from one end of the bundle to the other. The kinocilium is situated eccentrically at the margin of the bundle and has a 9+2 arrangement of microtubules. The hair projects above the surface of the epithelium, and the cells are enclosed in a gelatinous tube called a cupula. The cupula lies accessible and can be moved by the water or by mucus. The sensory cell is connected to the central nervous system by afferent and efferent nerve fibres. The cell is sensitive to deflection and is a mechano-receptor.

Water movements bend the protruding cupula, stimulating the hair cell by bending the attached cilia (sense hairs). Most teleosts and elasmobranchs have also developed lateral line canals, in which the neuromasts lie between canal pores open to the environment. The cupula of the canal neuromasts are sensitive to movements of the watery, endolymph fluid through the canal. Like the auditory hair cells, the neuromasts continuously send neural impulses to the brain.

The lateral line system is developed and used in various ways by fishes exhibiting different modes of life. Parts of the lateral line system are often specialized for detecting prey. Surface-oriented fishes such as killifish (*Fundulus*) sense the ripples from the water surface caused by struggling insects by ringing the short canal neuromast system on the flattened dorsal surface of their heads. Blind cavefish have rows of free neuromasts on the head, which stand out in ridges to aid the cavefish in locating the food so precisely that it can be ingested without sight.

Neuromast function can also be helpful to fish for stimuli other than localized water disturbances from distant objects. As the fish swim, they "push" some water in front of their heads, as a boat creates a bow wave. The amount of water pushed or displaced depends on the size of the fish and the shape of the head. These forward water displacements rebound from objects in front of the swimming fish. The neuromasts on the head apparently sensed the rebounds, allowing avoidance of the obstacle.

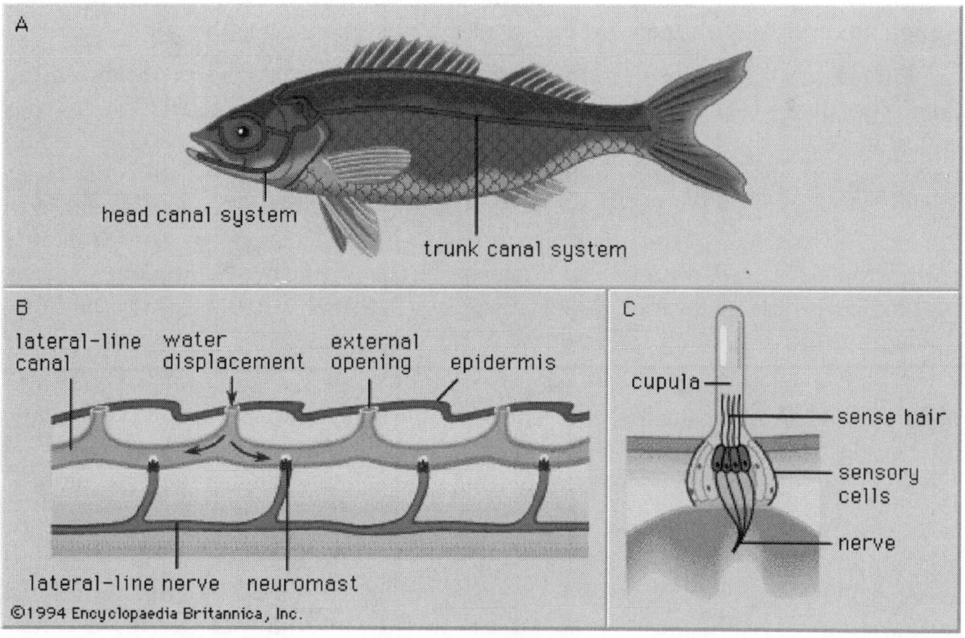

Fig 2.11 - Lateral line

3.2.3. Chemoreception

Odours and tastes are quite distinguishable to the great majority of terrestrial animals. Olfactory organs, stimulated by airborne molecules, are more sensitive and chemical-specific than gustatory organs, which are generally stimulated by contact with dilute solutions. In fish, both types of organs are sensitive to 'contact' stimuli of chemicals dissolved in water. Besides differences in sensitivity, the 2 sensory systems are also distinguished by the location of the sensory receptors. Olfactory organs reside in special pits; gustatory receptors are often more dispersed on external surfaces or in the mouth.

3.2.3.1. Olfaction

The sense of smell is important in almost all fishes. In fishes, the olfactory organs are still more sensitive of the 2 means of chemoreception. The olfactory or nasal organ of fishes is located on the dorsal surface of the snout. The olfactory receptors are usually located in olfactory pits, i.e., the nasal sac with incurrent and excurrent channels divided by a skin flap. Water is induced to flow through olfactory pits by movements of cilia within the pit by the muscular movement of the branchial pump, by swimming or by a combination of these. Odours are perceived when the dissolved chemical contacts the olfactory rosette, a multi-folded epithelium rich in receptor cells in the olfactory pit. Fishes relying heavily on olfactory cues have elongated rosettes with many receptor cells in elongated olfactory pits.

Olfactory stimuli are communicated to the olfactory lobe of the brain via the first cranial nerve. Species that rely heavily on olfactory information, such as anguillid eels, moray eels and sharks, display oversized olfactory lobes. Certain eels with tiny eyes depend mainly on their sense of smell to locate the food. A few fish, such as puffer fish, have significantly reduced lobes, consistent with the evolutionary loss of the olfactory organs and pits. These puffers probably rely entirely on sight for feeding. Odour also serves as an alarm system. Many fishes, especially various species of freshwater minnows, react with alarm to the body fluids produced by an injured member of their species.

Some olfactory stimuli are amino acids, bile acids, gonadal steroids and derivatives, prostaglandins and various alcohols, amines, carboxylic acids, nucleotides, and aromatic hydrocarbons.

Olfactory cues have been shown to be necessary for salmon to locate their natal stream once they are near the river mouth. Thus, salmon are thought to be imprinted with odours as pre-smolts and smolts migrate down to the rivers and streams to a larger body of water where they spend the majority of their adult life. In sharks and rays, the paired pits that lead into the internally folded nasal sac are usually on the ventral side of the snout and divided by a fold of skin into an inlet and an outlet. Studies on nurse sharks show a distinct gradient searching activity for food detection. The shark turns in that direction by detecting a stronger chemical concentration in the olfactory rosette on one side of its head.

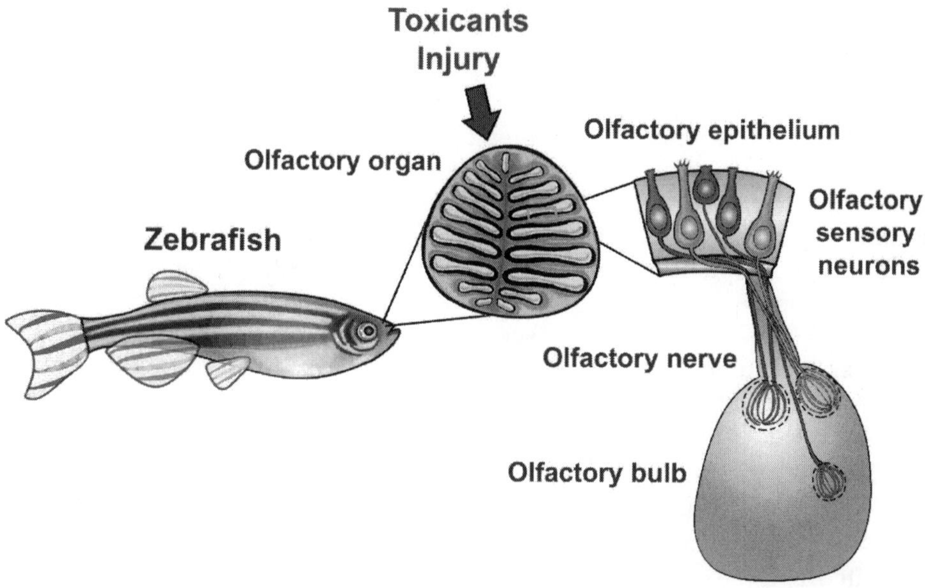

Fig 2.12 – Olfactory System

3.2.3.2. Taste or gustatory chemoreception

Taste or gustatory chemoreception is generally a "close range" sense in fishes, especially useful in identifying food and noxious substances. Whereas the olfactory receptors are localized in the nares (sense of smell), taste buds are commonly located on several exterior surfaces of the fish besides in the mouth.

Many fishes have a well-developed sense of taste, and tiny pit-like taste buds or organs are located not only within their mouth cavities but also over their heads and parts of their body. Taste perception has not been extensively studied in bony fishes. Some species can detect some sensations, such as salty, sweet, bitter, and acid stimuli. Taste may be responsible for the final acceptance or rejection of prey items. Bottom fishes such as catfishes have considerable numbers of taste receptors on their skin, fins and barbels. The barbels (whiskers) of catfishes, which often have poor vision, serve as supplementary taste organs, those around the mouth being actively used to search out food on the bottom. The high gustatory sensitivity of the barbels extends the usefulness of these chemoreceptors, which are a reasonable distance from the fish, presumably to aid in finding food in murky water. A similar distant touch sense might be inferred from the presence of taste buds on the free pelvic fin rays of some codfishes and *Trichogaster trichopterus*. Other structures bearing recognizable taste buds include the well-developed lips of some minnows and the suckers. Taste buds have also been described on the palatal organs in the buccal cavities of various cyprinids, catostomids (**Catostomidae** is the sucker family of the order Cypriniformes) and salmonids as well as on the gill rakers and arches primarily in freshwater fishes. Some species of naturally blind cave fishes are exceptionally well supplied with taste buds, often covering most of their body surface.

Gustatory information is transmitted to the brain of the vertebrates by 3 pairs of cranial nerves: the facial (VII), the glossopharyngeal (IX) and the vagus (X). Facial nerve branches, which innervate primarily the taste buds located on the external body surface, together with their central projections, constitute the extra-oral gustatory system. This system is implicated in the detection and localization of a food source. The glossopharyngeal and the vagal nerve branches innervate only those buds located in the oro-pharyngeal region and thus constitute the oro-pharyngeal taste system, which is essential for selective food ingestion. Branches of the facial cranial nerve VII innervate the cutaneous taste buds. **Cranial nerves** are nerves that emerge directly from the brain, in contrast to spinal nerves, which emerge from spinal cord segments. The sensory signals from the internal taste buds (e.g., those in the pharyngeal cavity and palate) are transmitted to the glossopharyngeal (IX) and vagus (X) nerves.

In species having a palatal organ, the total number of taste buds is considerably augmented, and the terminal centres in the visceral sensory column of the medulla are correspondingly enlarged as "vagal lobes". Just as the olfactory lobes of the brain are grossly enlarged in fishes relying heavily on olfactory sensory information (e.g. Atlantic eels, sharks), the facial or vagal lobes of the medulla display a marked enlargement in species that find food by the sense of taste.

The functional importance of taste receptors to catfishes has been demonstrated by surgical removal or sectioning of optic and olfactory senses but does not prevent catfish from swimming to a food source in still water. When taste reception is surgically blocked on one side of the catfish, it continuously circles toward the side of intact reception until, eventually, the food is found. Surgical removal of the facial lobes of the medulla prevented catfish from locating and ingesting food. In contrast, removing the vagal lobes did not inhibit finding food or taking it into its mouth but prevented swallowing, which is controlled via the glossopharyngeal and vagal nerves. Interestingly, sharks and rays show no sensory elaboration for taste functions. Elasmobranchs rely primarily on olfaction, vision and electro-reception to locate their food.

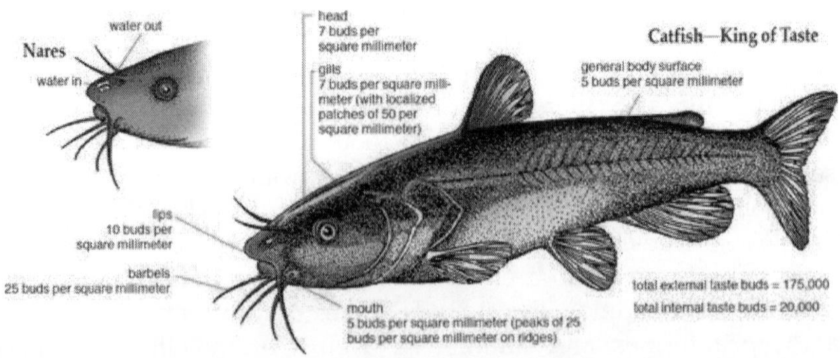

Fig 2.13 - Gustatory chemoreception

3.2.4. Electro-receptor Organs

They are highly sensitive to the electric field and are used for electrocommunication, electrolocation and social signalling. The electro-receptors show considerable variation in their morphology in different species and are distributed all over the skin of the fish. In fishes, electro-reception is found among all non-teleost taxa except Myxiniformes and Holostei. Two kinds of electro-receptors are found in fishes.

3.2.4.1. Ampullary organ (Ampullae of Lorenzini)

Cartilaginous fish related to sharks and rays possess a unique electrosensitive sensory organ in the skin around the head known as the Ampullae of Lorenzini. This organ is responsible for detecting weak electric field changes caused by the muscle contractions of the prey. They respond to signals of low frequencies up to 50 Hz

3.2.4.2. Tuberous organs

They are high-frequency receptors restricted to electric fishes that produce their own high-frequency electric signals.

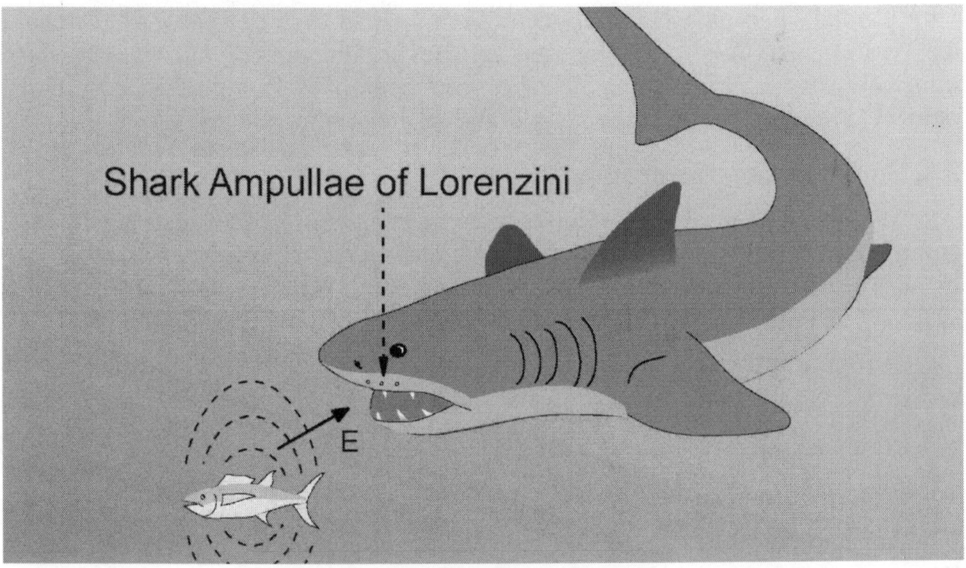

Fig 2.14 - Ampullae of Lorenzini

Function of Electro-receptor

The electro-receptors enable the fish to locate objects at a distance, as the fish possessing them are far more sensitive to the electric field than those without them. The fish become aware of any distortion in the electric field caused by another animal or inanimate object.

Fig 2.15 - Electroreceptors

3.2.5. Other senses (touch, pain, and special senses)

A fish or other vertebrate seldom has to rely on a single type of sensory information to determine the nature of the environment around it. Like most other animals, fishes have many touch receptors over their body surface. Pain and temperature receptors are also present in fishes and presumably produce the same information to fish as to humans. Fishes react negatively to stimuli that would be painful to human beings, suggesting that they feel a sensation of pain.

3.3. Sense organs in prawn

All the specialized organs that receive physical and chemical environmental stimuli are called sense or sensory organs. The sense organs are associated with the nervous system and transmit the environmental stimuli to the central nervous system. The sense organs of a prawn consists of the following organs:

1. Compound Eye
2. Statocysts
3. Other sense organ

- Tangoreceptors
- Chemoreceptors
- Proprioreceptors

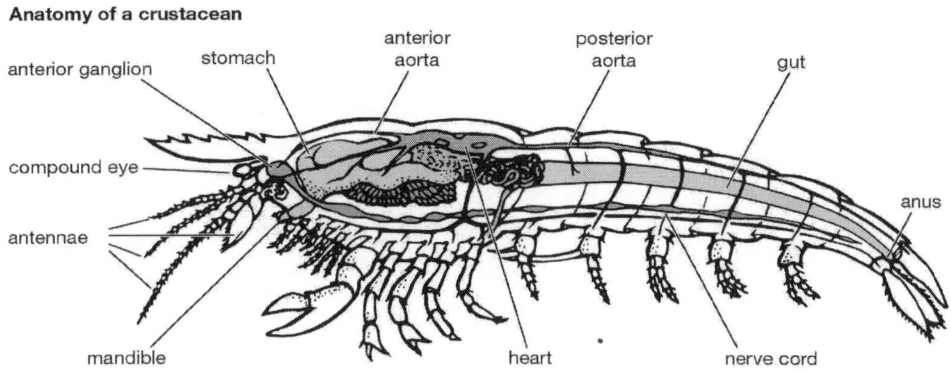

Fig 2.16 - Sense organs in prawn

3.3.1. Compound Eye

Prawn has a pair of black and hemispherical eyes. The eye is located at the base of the rostrum and connected by two jointed, short, movable stalk. Each eye is composed of a large number (about 2500 no.) of independent visual elements or units called ommatidia, which is why the eye is called a compound eye. Each of the ommatidia is arranged radially and has a similar structure.

Mosaic vision

Each ommatidium can produce a separate image of a small part of the object seen. Therefore, the object viewed consists of several tiny dark and light spots, so the image forms a mosaic. Because of its similarity to mosaic artwork, this type of vision is called mosaic vision. Prawns can adjust their eyes to form both images according to the prevailing intensity of light. The optic nerve carries impulses to the brain, where the interpretation and registration of images occur.

3.3.2. Statocysts

Statocysts act as organs of orientation and equilibrium. Statocysts are a pair of tiny, white bead-like sacs that lie inside the basal segment or precoxa of each antennule to its dorsal wall. It opens through a minute concave depression of precoxa called statocystic aperture. A small statocystic branch of the antinnular nerve supplies to the statocyst. The oval-shaped statocystic cavity is filled with minute sand particles and lined by a number of elongated sensory hairs of receptor setae. Each receptor setae connected by a statocystic nerve. Each receptor setae consists of a swollen base and a long tapering shaft pointed towards the centre.

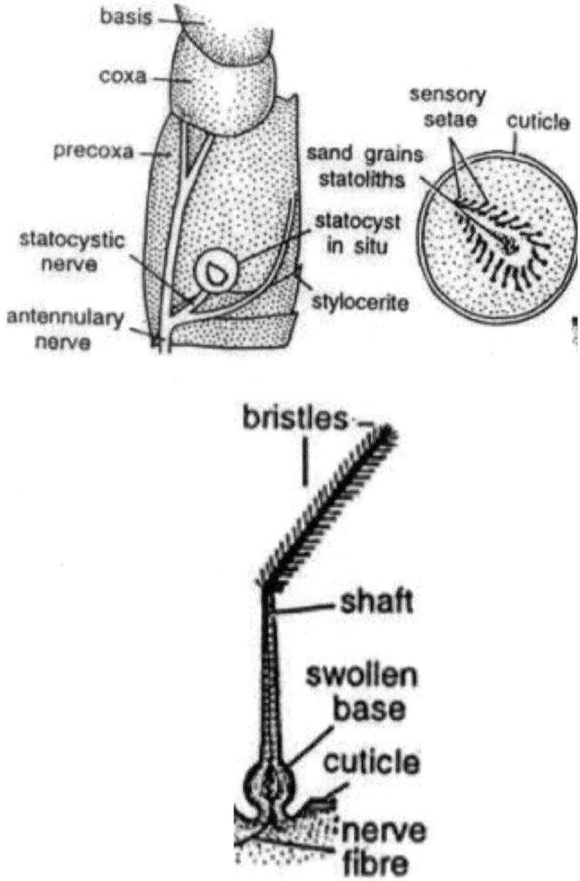

Fig 2.17 - Statocysts

3.3.3. Other sense organs

1. **Tango-receptor:** Prawns are sensitive to touch through tactile organs or tango-receptors. Elongated feelers of both the antennae are said to be tactile in function.
2. **Tango-receptor** Chemo-receptors or olfactory organs respond to chemical stimuli. They present on mouth parts, flagella of antennules and the inner wall of the gill chamber. The nerve fibre from the olfactory branch of the antennulary nerve is connected to each seta.
3. **Proprio-receptor:** These occur internally throughout the body. They receive internal stimuli such as related to posture and muscular function.

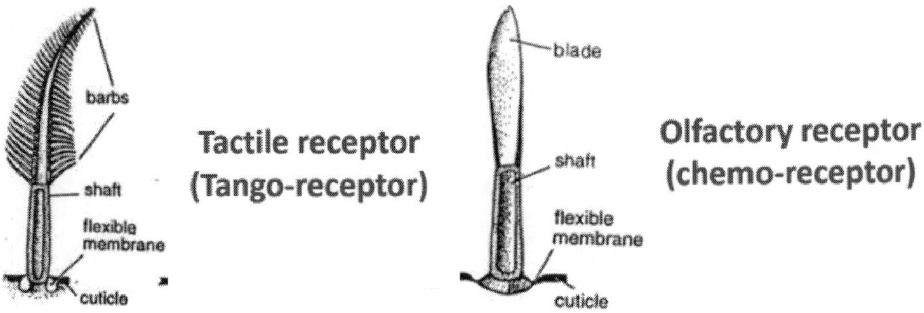

Fig 2.18 - Tango-receptor and Chemo-receptor

UNIT III

CHAPTER 1

ENERGY AND NUTRIENT STATUS OF FOOD

1.1. Introduction

Energy is defined as the capacity to do work and is derived by animals through the catabolism of dietary carbohydrates, lipids and proteins within the body. Many forms of energy exist in nature, i.e., radiant, chemical, mechanical, heat, and electrical energy. Energy is essential for maintaining life processes such as cellular metabolism, growth, reproduction, and physical activity. Life on earth depends on radiant solar energy and its subsequent fixation and conversion by green plants during photosynthesis into stored chemical energy (i.e., carbohydrates) for use as an energy source by plants and animals.

1.2. Laws of thermodynamics

All forms of energy are inter-convertible and obey the laws of thermodynamics. The first law of thermodynamics states that energy may be transformed from one form into another but can never be created or destroyed.

Solar energy → Chemical energy + Heat energy

The second law of thermodynamics states that no transformation of energy will occur unless energy is degraded from a concentrated form to a less concentrated or more dispersed form, and further that no transformation is 100% efficient.

Solar energy (high energy form) → Chemical energy (low energy form)

1.3. Energy units

Energy is usually expressed in terms of heat units since all forms of energy are convertible into heat energy. The primary heat unit normally used is the calorie. One calorie is defined as the amount of heat required to raise the temperature of one gram of water by one degree centigrade. Since, for many purposes, the calorie (cal) is too small a unit of measurement, the kilocalorie (kcal) is often used; 1 kcal

= 1000 cal. In many scientific studies, the calorie is now being replaced by the joule (J) as the unit of energy; 4.184 J = 1 cal.

1.4. Energy Status of food

Carbohydrates give average gross energy values of 4.2 kcal or 17.6 kJ per gram

Fat gives 9.4 kcal or 39.4 kJ per gram

Protein gives 5.65 kcal or 23.7 kJ per gram

1.5. Energy metabolism

Metabolism is usually the most significant part of the energy budget and has a central role in physiological traits like growth and energy storage. Metabolism is the combination of anabolism (synthesis) and catabolism (breakdown). Energy metabolism is concerned with the catabolism and oxidation of carbohydrates, lipids and proteins within the animal body and the consequent release and use of the liberated energy as work for the maintenance of the life process.

1.6. ATP (Energy Currency of the cell)

The free energy liberated from the catabolism and oxidation of the major food nutrients is not utilized directly by the animal but instead is trapped in the chemical form of the energy-rich phosphorus bond of adenosine triphosphate (ATP). It is ATP, the principal driving force in the energy-requiring biochemical processes of life. Aerobic respiration consumes the glucose and produces 30 or 32 moles of ATP, equivalent to 262.8 kcal. Anaerobic respiration produces only 2 moles, equal to 21.9 kcal.

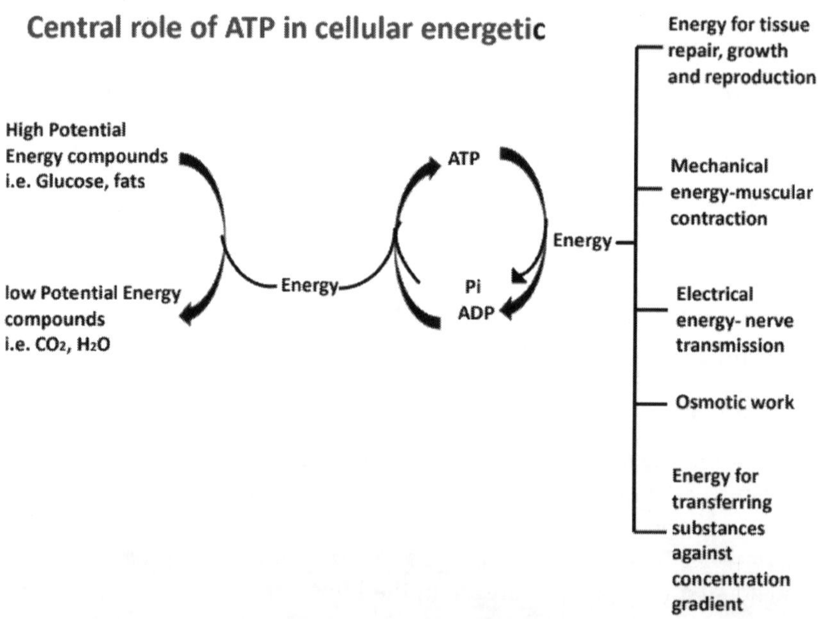

Fig 3.1 – Central role of ATP

1.7. Extraction of Energy from Foodstuffs

In the first stage, large molecules in food are broken down into smaller units. Proteins are hydrolyzed to their 20 kinds of constituent amino acids, polysaccharides are hydrolyzed to simple sugars such as glucose, and fats are hydrolyzed to glycerol and fatty acids.

In the second stage, these numerous small molecules are degraded to a few simple units that play a central role in metabolism. Most of them sugars, fatty acids, glycerol, and several amino acids are converted into the acetyl unit of acetyl CoA.

In the third stage, ATP is produced from the complete oxidation of the acetyl unit of acetyl CoA. The third stage consists of the citric acid cycle and oxidative phosphorylation, the final common pathways in the oxidation of fuel molecules. Four pairs of electrons are transferred (three to NAD+ and one to FAD) for each oxidized acetyl group. Then, a proton gradient is generated as electrons flow from the reduced forms of these carriers to O_2, and this gradient is used to synthesize ATP.

1.8. Pathways of cellular metabolism

The enzymatic degradation of each of the major energy-yielding nutrients of cells (CHO, fats and proteins) is produced stepwise through several consecutive enzymatic reactions. The large bio-molecules during digestion have been reduced to their building blocks, viz. hexoses/pentoses, amino acids and fatty acids and glycerol. During the first phase of catabolism, these substances (building blocks) are converted into smaller and simpler molecules. Thus, hexoses, pentoses, amino acids and glycerol are degraded to a single 3-carbon intermediate pyruvate, which is then converted into a single 2-carbon unit, the acetyl group of acetyl CoA. In the next phase, Acetyl CoA is fed into the citric acid cycle, the final common pathway by which most nutrients are ultimately oxidized to CO_2, water and NH_3 as the end productions.

During the initial place of catabolism, one glucose molecule (6C) is degraded by 10 successive reactions to yield 2 moles of pyruvic acid (3C). This part of the metabolism is called Glycolysis. Glycolysis is a central pathway in the catabolism. Oxidation of pyruvic acid (and further reactions) required O_2. When some muscles work vigorously, the pyruvic acid cannot be oxidized due to the lack of O_2 under such conditions, and the pyruvic acid is converted into lactic acid. In the anaerobic micro-organisms and some fungi, the pyruvate is converted into ethyl alcohol. This process is called (alcoholic) fermentation. The amount of energy released from glycolysis will be far less than that released from complete oxidation of the glucose molecule.

The pyruvic acid formed is oxidized to form the acetyl group of acetyl CoA. The acetyl group is then oxidized entirely to CO_2 and water by the citric acid cycle, TCA cycle, or Krebs cycle with the intervention of molecular oxygen. In contrast, glycolysis occurs in the cytoplasm. Krebs cycle and electron transport system occur in the mitochondria.

Breakdown of energy molecules into smaller unit

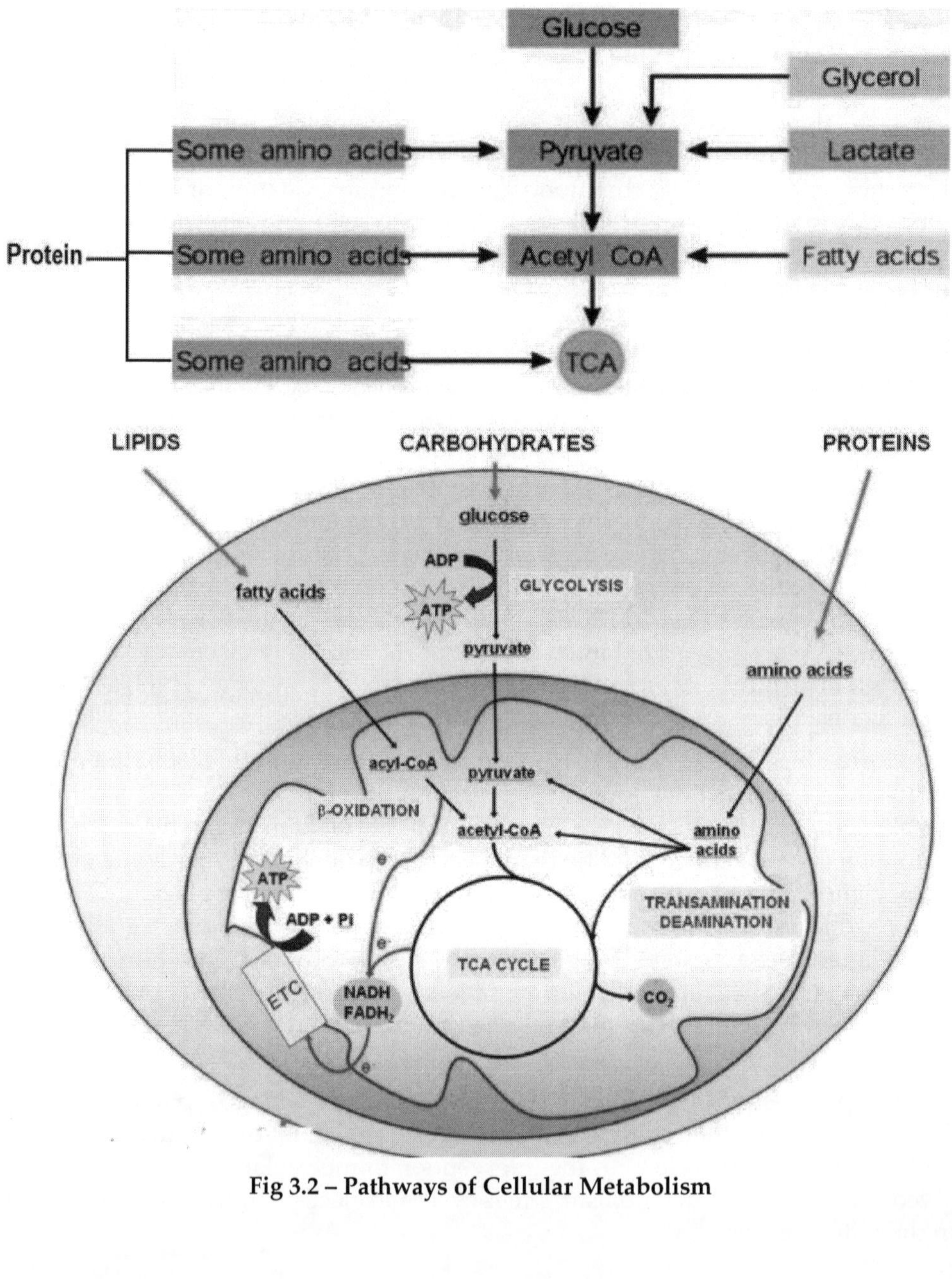

Fig 3.2 – Pathways of Cellular Metabolism

1.9. Mechanisms of ATP Synthesis

There are two mechanisms of ATP synthesis:

1. **Oxidative phosphorylation** is the process of synthesizing ATP from ADP and inorganic phosphate (Pi), which occurs in the mitochondrion.
2. **Substrate-level phosphorylation,** in which ATP is synthesized by transferring high-energy phosphoryl groups from high-energy compounds to ADP. It occurs in the mitochondrion during the tricarboxylic acid (TCA) cycle and in the cytoplasm during glycolysis.

1.10. Energy balance and dietary requirement

Only a portion of the total chemical energy contained within the food ingested is available to the animal (i.e., net energy) for maintenance, activity and growth; 45% of the ingested food energy is lost as undigested food (faeces), metabolic excretion and heat.

The energy balance equation

$$C = P + R + U + F$$

where C (for consumption) is the gross energy content of the food ingested, P is the energy utilized in growth materials (production), R is the net loss of energy as heat (for respiration), U (urinary loss) is the energy lost in nitrogenous excretory products, and F is the energy lost in the faeces.

1.10.1. Digestive System

The structural components of a fish's digestive system include the mouth, teeth and gill rackers, oesophagus, stomach, pylorus, pyloric caeca, pancreatic tissue (**exocrine** and **endocrine**), liver, gall bladder, intestine and anus. Not all components are present in all fish. A fish's digestive system is adapted to their food habits. In predatory (carnivorous) fishes, the mouth is usually large for engulfing prey as a whole or in large chunks, and teeth are present on the jaws (*e.g.*, maxillary and dentary) and tongue for grasping live prey. Gill rackers are short in carnivorous fish, and pharyngeal teeth are short and pointed to move prey down the throat. In omnivorous and **planktivorous** fishes, the mouth is more petite and is usually devoid of teeth except for pharyngeal teeth that may be blunt and flat (molariform) for grinding or sharp and long for shredding. Gill rackers in these fish are very fine in structure to prevent the escape across the gills of small food particles.

The oesophagus, which is lubricated by mucus, leads to the stomach. In carnivorous fish, the stomach is muscular and elastic for holding large prey items, while in omnivorous and planktivorous fishes, the stomach, if present at all, is small because a more or less constant stream of small food particles can flow directly into the intestine. The pylorus is a sphincter that prevents premature movement of the food **bolus** out of the stomach. Around the pylorus, many fish have out-pocketings called pyloric caeca. The role of pyloric caeca in digestion has been widely speculated. Histologically, they are very similar to the intestine. Around the pyloric caeca (or in that same area in fish lacking them) is pancreatic tissue. In fish the pancreas is usually diffuse, not a discrete body. As in all vertebrates, the pancreas has two digestive functions. It is the source of:

1. Exocrine secretion of digestive enzymes - **proteases, lipases,** and **amylases** into the intestine and
2. Endocrine secretion of the hormones - insulin and glucagon

The liver in fish produces bile, which is stored in the gall bladder until a bolus passes the stomach. At that time, the bile is expelled into the intestine. Bile contains waste products from liver activity that pass out of the fish in the faeces. Bile has a digestive function of emulsifying lipids, significantly increasing their absorption in the intestine. The liver is vital in the **anabolism** and **catabolism** of amino acids absorbed during digestion and is also the site of storage of food energy in the form of **glycogen**. Most absorption of nutrients occurs in the intestine. The large protein, fat, and starch molecules in food are broken into smaller molecules by gastric acid, and digestive enzymes move by diffusion or active transport (ATPase pumps) into the network of capillaries surrounding the gut. The intestine is lined with finger-like out-pocketings (villi) that significantly increase the surface area for absorption. The intestinal wall, while not as active as the exocrine pancreas, can also secrete digestive enzymes. The **alimentary** canal is typically a short, S-shaped tube in carnivorous fish, but can be long and convoluted in omnivorous fish to aid in the absorption of less digestible plant material.

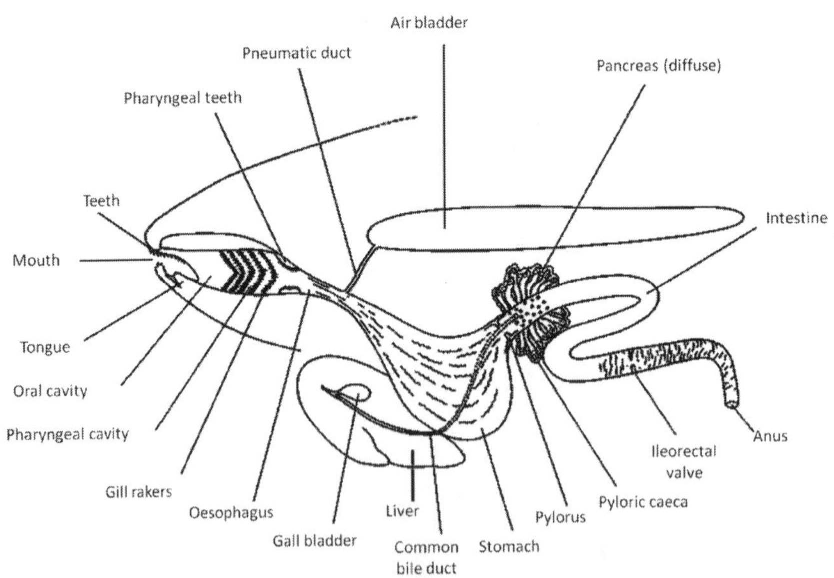

Fig 3.3 - Digestive System

1.10.2. Oxygen Effects

A fish has a constant metabolic rate over a broad range of dissolved oxygen (DO) concentrations at a given temperature and activity level. However, there is a critical point of DO below which a fish cannot extract enough oxygen from the water to support its metabolic rate. At this point, a fish begins to accumulate oxygen debt. It must generate some of its ATP from anaerobic sources and lactate, which cannot be further oxidized and accumulate in the tissues. A fish cannot live indefinitely below the critical point. At some point, as DO continues to fall, the fish cannot sustain life anaerobically and die from asphyxiation.

1.10.3. Activity Effects

In all animals, a higher rate of physical activity (exercise) oxidizes more calories. In fish, whose metabolic rate is affected by the oxygen content of the water, the oxygen effect interacts with the activity effect. As the rate of activity increases, the fish needs more oxygen in the water to sustain aerobic metabolism.

1.10.4. Species Effects

At any given temperature, oxygen and activity level, two different species may have different metabolic rates. For example, a carp and a trout have different haemoglobin oxygen dissociation curves and tissue tolerances for hypoxia. This allows the carp to tolerate lower DO conditions without going into oxygen debt (lower critical point) and trout to sustain a higher activity rate (higher critical point). It is an evolutionary trade-off that adapts species to different habitats and behaviours.

1.10.5. Effects of the Aquatic Habitat

All animals create **ammonia, which is produced by** the catabolism of proteins. Ammonia is toxic, and its accumulation in the body can not be tolerated. Mammals use energy to turn ammonia into less toxic urea in the liver. Urea is water soluble, and mammals excrete it in the urine. In order to reduce the weight of water for flight, birds create another high-energy nitrogenous waste product, uric acid, which is excreted with the faeces. Since fish constantly exchange with water at the gill, they can excrete ammonia directly, without conversion, as soon as it is created. Mammals and birds pay a caloric penalty for eating protein. When carbohydrates or fat are metabolized, about 5% of the assimilated energy is used to change the structure of the food molecules into forms the body can use, and the remaining 95% can be used for maintenance and growth. When mammals and birds metabolize protein, however, the penalty increases to about 15% because energy must be used to build urea or uric acid, and these energy-containing substances are excreted from the body and lost.

1.10.6. Nitrogenous Compounds

Since fish do not have to expend energy to change ammonia to urea or uric acid, they do not have to pay the energy expense to use protein for food as birds or mammals do. Fish pay about a 5% penalty whether protein is part of the meal or not. Since fish do not pay a penalty for eating protein, they can direct more energy in food towards production.

1.10.7. Effects of Poikilothermy

The other way fish are more efficient in converting food energy into production is that fish do not use energy to maintain a body temperature above ambient temperature. Mammals and birds use energy constantly to keep their body heat up. In cold temperatures, this is relatively substantial. So, poikilothermy and a lack of a penalty for protein consumption are why fish can have high feed conversion ratios.

1.10.8. Effects of Carnivory

Regarding both biomass and diversity, the terrestrial environment is dominated by herbivorous animals. Carnivores are in the minority compared to herbivores. In the aquatic environment, fish are the dominant vertebrate, and most are carnivorous. Herbivorous fish exist but are quite rare compared to carnivorous and omnivorous fish. Moreover, there are no **ruminant** fish, while many terrestrial mammals are ruminants.

In the U.S., most fish culture is directed toward intensely carnivorous trout and salmon or somewhat omnivorous channel catfish. Fish that are more fully omnivorous, *e.g.*, tilapia, are less commonly produced in the U.S. but commonly imported from tropical countries. In salmonid rations, the food is fed with a very high percentage of protein (30 to 40%). Not only is the percentage of protein high, but the majority of this protein must come from animal sources. This is because the amino acid balance is different between animal and plant protein, *i.e.*, all the essential amino acids are present in plant protein, and they are all present in animal protein. However, the percentage of individual amino acids tends to differ between animal and plant sources, and less total protein needs to be fed if the amino acid profile of the feed is similar to the animal's body.

Moreover, carnivorous fish like trout do not tolerate much carbohydrate in their diet (it cannot exceed 20% starch and sugar), and it is difficult to feed enough plant protein to meet their amino acid requirements without exceeding this limit since plants are so rich in carbohydrates. Consequently, the best protein source for carnivorous fish is fish meal, which means feeding fish flesh to grow fish flesh. This is expensive and not a very ecologically sound way of growing food. An irony of carnivorous fish aquaculture is that it increases the use of wild ocean fish. Growing trout and salmon is like raising lions and tigers for food. In the recent years before scientific ration formulation allowed a complete trout ration to be made from dry feedstuffs, trout were successfully raised on the ground beef liver. Channel catfish are more tolerant of plant material in their diet, but most catfish rations include a sizable percentage of fish meal.

1.11. Least Cost Ration Formulation

The goal of the production of animal nutrition is the use of a scientifically formulated least-cost ratio. This can only be done if the animal's nutritional requirements have been determined. The precise needs for crude protein, fat, carbohydrates, vitamins, and individual amino acids were not determined for any fish until the 1950s when the necessary research for rainbow trout was completed. At this time, complete nutritional requirements are only known for salmon, channel catfish and rainbow trout, though this will probably be determined soon for tilapia. Once the nutritional requirements are known, a ration can be formulated that meets those requirements by combining various commercially available feedstuffs (soybean meal, corn meal, fish meal, feather meal, etc.) in the combination of ingredients that meet the fish's nutritional needs at minimum cost. The formula is then mixed and extruded into pellets. Such dry diets have a reasonably long shelf life and are easy to store and handle.

CHAPTER 2

NITROGEN BALANCE

2.1. Ammonia production

Ammonia is the primary end product of Nitrogen metabolism in almost all osteichthyes, agnathans, and freshwater chondrichthyes, but less critical than urea in marine chondrichthyes, dipnoans during estivation, and very few teleosts living in extreme environment.

The largest source of ammonia is catabolism of dietary or structural protein. Essential amino acids in fish are the same as those in mammals: arginine, histidine, isoleucine, leucine, lysine, methionine, phenylalanine, threonine, tryptophan, valine, and all others can be synthesized. Fish have a remarkable capacity to utilize amino acids both as metabolic fuel and as precursors for protein, lipid, and carbohydrate synthesis. In actively growing fish fed a balanced diet, greater than 50% of dietary N-content is incorporated into structural growth, and most of the remainder is used for energy production. Dietary carbohydrates do not appear to be a crucial aerobic fuel, but there is a high capacity for gluconeogenesis from exogenous amino acids. However, dietary lipids are undoubtedly essential, especially in carnivores. If the diet is deficient in lipids, then a more significant proportion of dietary protein is metabolized for energy or deaminated for conversion to fat and carbohydrate, more ammonia is excreted, and a lower percentage of dietary N is retained for growth.

All fish exhibit a marked ability to consume their own structural protein during prolonged starvation, but the size of the endogenous fraction as a proportion of metabolic rate varies depending on the extent of fat reserves. Glycogen reserves tend to be spared during starvation, probably because of their survival value as an anaerobic fuel during burst exercise and hypoxia. The other significant source of ammonia is the deamination of adenylates.

When fish feed and grow, absorbed amino acids in excess of those needed for protein synthesis are deaminated and then oxidized in Kreb's cycle or converted to fat and carbohydrates. When fish are starving or energetic expenditure exceeds intake, amino acids released by muscle proteolysis are similarly deaminated. The most important pathway is thought to be the transamination system. Various aminotransferases transfer the amino group of L-amino acids to alpha-ketoglutarate to form glutamate, which is subsequently deaminated by glutamate dehydrogenase. This occurs in many tissues, including muscle, gill, and kidney, but enzyme activities are generally highest in the liver. The other major pathway is probably the direct hydrolysis of the amide groups on glutamine and asparagine, catalyzed by glutaminase and asparaginase. These enzymes occur in the liver, kidney and gills, and at least glutaminase is present in red and white muscle.

The purine nucleotide cycle has been suggested as another deamination route. Here, the amino group on glutamate is transferred to oxaloacetate by aspartate aminotransferase to form aspartate and regenerate alpha-ketoglutarate. Aspartate then reacts with inosine monophosphate (IMP) and guanosine triphosphate (GTP) to form adenylosuccinate, mediated by adenylosuccinate synthetase, followed by conversion to fumarate and adenosine monophosphate (AMP), mediated by adenylosuccinate lyase. The deamination of AMP by AMP-deaminase liberates ammonia and regenerates IMP, thereby completing the cycle. In addition, fumarate may be converted to malate by fumarase and then back to oxaloacetate by malate dehydrogenase, thereby regenerating the recipient of the amino group. While the enzymes of this cycle are present in the liver, it does not appear to be the main hepatic pathway of amino acid deamination.

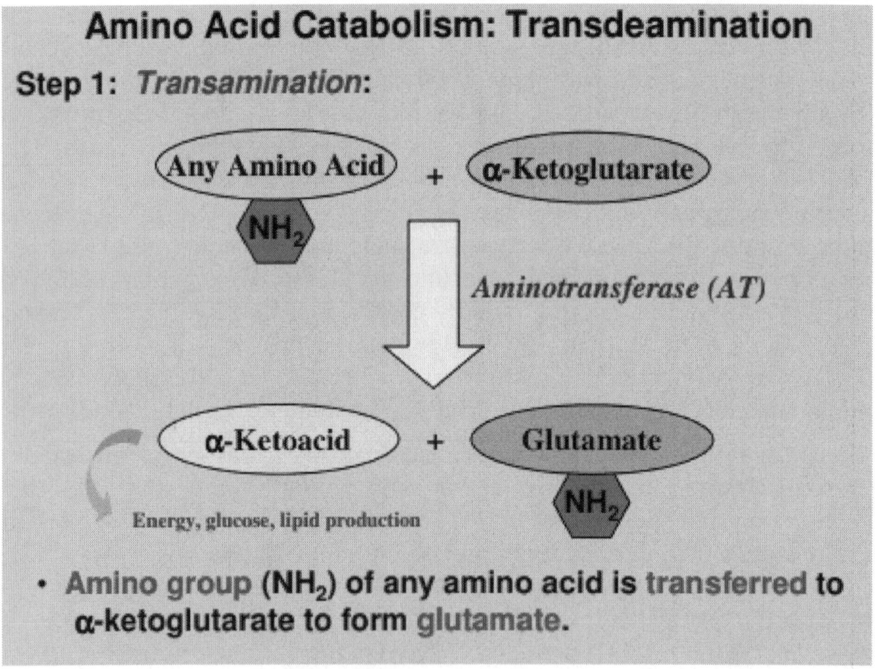

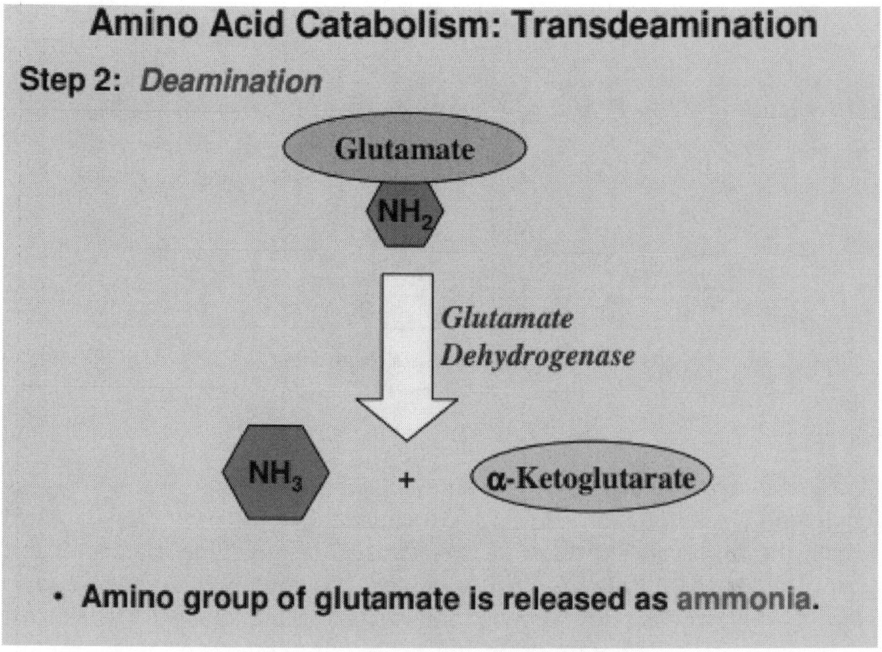

Fig 3.4 – Ammonia production

2.2. Ammonia excretion

It is sufficient to note that evidence exists for at least four mechanisms of ammonia extrusion: (1) NH_3 diffusion along a PNH_3 gradient from blood to water; (2) electroneutral Na^+ / NH_4^+ exchange located on the apical membranes of branchial epithelial cells; (3) a similarly located electroneutral H^+/NH_4^+ exchange; and (4) NH_4^+ diffusion along an electrochemical gradient from blood to water. If there is any consensus, it would be that the first two are the principal mechanisms under most normal conditions and that mechanism (4) is only essential for teleosts in seawater, where cationic permeability is much higher than in freshwater. The term "Na^+/ NH_4^+ exchange" is used here in a general sense, recognizing that its exact nature remains uncertain. Thus, NH_3 diffusion stoichiometrically tied to a classic Na^+ / H^+ exchange or a primary W-ATPase/Na^+ channel by a local pH gradient would also manifest as electroneutral Na^+/NH_4^+ exchange. At least in marine fish, there is evidence of two additional mechanisms for entry through the basolateral membrane: (5) substitution of NH_4^+ for K^+ on basolateral Na^+/K^+ATPase in teleosts and (6) substitution of NH_4^+ for K^+ on basolateral Na^+, K^+, 2Cl cotransporters in elasmobranchs. Their acid-base consequences will depend on whether NH_4^+ or NH_3 leaves across the apical membrane

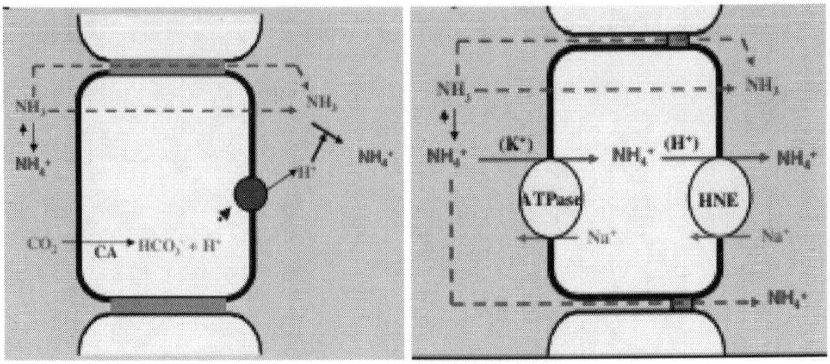

Fig 3.5 - Ammonia excretion

2.3. Urea

Urea is less toxic than ammonia because it exists in a single form, a non-charged molecule and does not exert direct electrochemical, acid-base or gaseous effects. However, the high concentration of urea causes destabilizing effects on the cell macromolecules (e.g., Proteins). An organism accumulating urea to high levels for osmoregulation has adopted a "counteracting solute strategy" in which high concentrations of TMAO, with opposing (stabilizing) effects on macromolecules, accumulate in tandem.

The production of urea takes place through the Ornithine urea cycle. It occurs in elasmobranchs, coelacanths and a few teleostean fish that survive air exposure or alkaline environments. It requires energy. It occurs primarily in the liver.

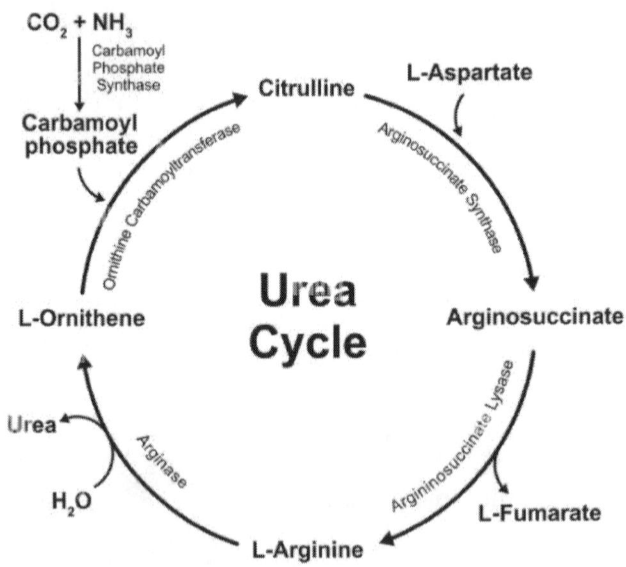

Fig 3.6 – Urea cycle

2.3.1. Transportation of Urea

Urea can diffuse through the lipid bilayer of the plasma membrane, so some urea is excreted by simple diffusion through the plasma membrane of hepatocytes and gill cells. Most of the urea is transported and excreted out from the body with the help of the urea transporter (transporter protein) present on the membrane of hepatocytes, RBCs and Gill cells.

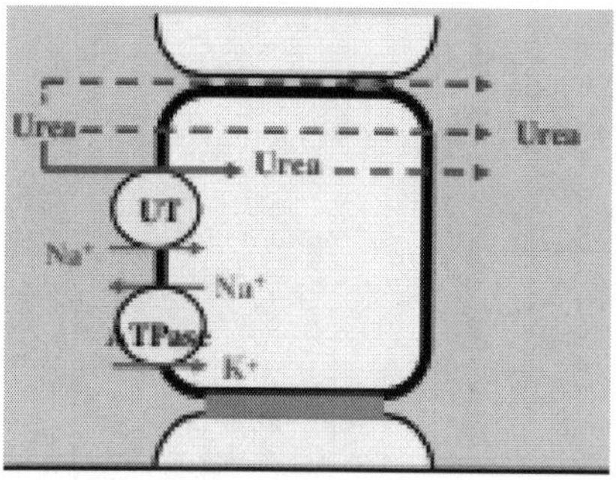

Fig 3.7 - Urea transportation

2.3.2. Urea Retention in Elasmobranch Gill

Minimal passive diffusion of urea into water transcellularly due to high cholesterol: phospholipid basolateral membrane. Some excretion is paracellular via "leaky" junctions only. Basolateral Na^+-K^+-ATPases create a gradient of low Na+ in the gill epithelium and high Na+ in the blood. Na^+ diffuses down its concentration gradient via a urea transporter (UT), transporting urea back into the blood.

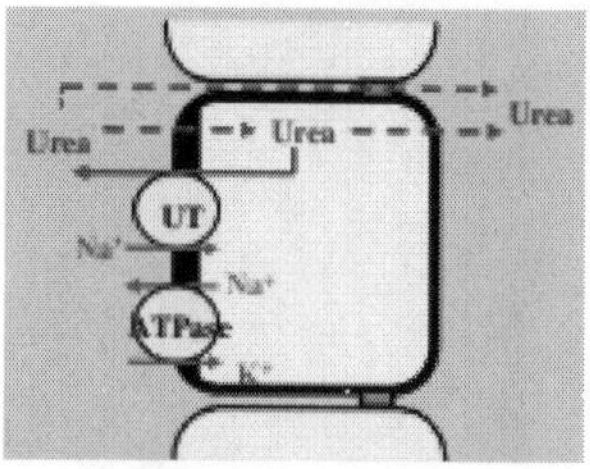

Fig 3.8 - Urea Retention in Elasmobranch

CHAPTER 3

METABOLISM

3.1. Digestion

Digestion is the process by which ingested materials are reduced to molecules of small enough size or other appropriate characteristics for absorption, i.e., passage through the gut wall into the bloodstream. This generally means that proteins are hydrolyzed to amino acids or polypeptide chains of a few amino acids, digestible carbohydrates to simple sugars and lipids to fatty acids and glycerol. Materials not absorbed are, by definition, indigestible and are eventually voided as faeces. Digestibility ranges from 100 percent for glucose to as little as 5 percent for raw starch or 5-15 percent for plant material containing mostly cellulose (plant fibre). The digestibility of most natural proteins and lipids ranges from 80 to 90 percent.

Digestion refers to the process by which ingested food materials get converted into forms of energy that they can utilize for activity, maintenance and growth.

It should be borne in mind that the process of digestion also requires energy. In fish digestion takes place by 2 processes.

1. Enzyme activity
2. Secretion of acids

Digestion starts when food is taken into the mouth and is completed when faecal matter is excreted from the anus, i.e., during a process of physical and chemical breakdown, ingested food items are reduced to their constituent molecules which are then absorbed in solution through the gut wall into the blood circulatory system. The food items are essentially broken down into smaller parts that can be absorbed and reconstituted.

Fishes have evolved various morphological and chemical adaptations for the digestion of food taken in. Once the food is taken in the mouth, it is pushed into the oesophagus. The oesophagus is a narrow tube that is highly elastic and contains many mucous layers. It functions as a lubricating transit for the passage of food material. From the oesophagus, the food is pushed into the lower gut. This lower gut is modified into the stomach in some fish. The primary function of the stomach is to store ingested food material, which is seen in carnivorous fish.

3.2. Digestion of proteins

The fishes, which possess stomach are generally carnivorous and secrete pepsin enzyme from gastric mucosa. The pepsin is a protease enzyme that can break down protein. HCl is secreted by the gastric mucosa in carnivorous fishes, creating a low pH because the enzymes are active in an acidic medium (pH 2). The trypsin enzyme is present in the pancreas extract of some elasmobranchs. The trypsin is secreted by the pancreas. The inactive form of this enzyme, trypsinogen, is known as zymogen. It is to be converted into an active enzyme, i.e., trypsin, by an enzyme enterokinase. The intestine of fish exclusively secretes the enterokinase enzyme. In the Cyprinids, stomach-less fish, protease compensation is supplemented by some intestinal enzyme called erypsin. The intestine secretes amino peptidases. They act on terminal amino acids called exopeptidases, and those acting on central bonds are called endopeptidases.

The enzymes which break down the carbohydrates in the gut of fishes are carbohydrases. They are amylase, lactase, saccharases/sucrease and cellulase. The most important enzyme is amylase, which acts on starch and breaks down to maltose and then to glucose by digestion. The starch is digested into glucose by amylase and maltase.

$$\text{Starch} \xrightarrow{\text{Amylase}} \text{Maltose} \xrightarrow{\text{Maltase}} \text{Glucose}$$

In those fishes in which sucrose has been reported, the effect is as follows:

$$\text{Sucrose} \xrightarrow{\text{Sucrase}} \text{Glucose} + \text{Fructose}$$

The amylase is secreted from the pancreas in carnivorous fishes, but in herbivorous fishes, this enzyme is reported from the whole gastrointestinal tract and the pancreas. Amylase is present throughout the alimentary tract of herbivorous fish like tilapia.

Blood glucose is converted, with the aid of insulin, to muscle glycogen. Although precise details are wanting, excess glucose enters the blood from the digestive tract, and the surplus is converted to glycogen in the liver. The fishes possess endo-commensal bacteria containing an enzyme, cellulase, which breaks down the cellulose plant material instead of passing it out through faeces.

3.3. Fat Digestion

The lipids are organic substances insoluble in water but soluble in organic solvents like chloroform, ether and benzene. They form essential dietary constituents due to their high caloric value due to the presence of fat-soluble vitamins and essential fatty acids. The main enzyme that acts on this lipid is lipase. This enzyme is found in the pancreas and mucosa of fishes. Lipase converts the fats into fatty acids and glycerols.

3.4. Absorption

Absorption is the body's process of taking up compounds through the stomach lining or intestinal walls. The indigested food material is thrown out as excreta or faecal matter. The amount of food the fish absorbs is determined by the difference between the amount ingested and excreted.

3.5. Metabolism

Nutrients consumed by fish are digested in the gut, absorbed by the gut lining, and appear in the bloodstream as their component molecules.

These molecules circulate in the body and are taken up by various tissues, where they are subjected to several different chemical reactions. **Metabolism** refers to the physical and chemical processes that occur inside the body's cells, which are necessary to maintain life. In metabolism, some substances are broken down to yield energy for vital processes, while others are synthesized as necessary for life. The transformation of the macronutrients such as **carbohydrates**, fats, and proteins in food to energy and other physiological processes is part of the **metabolic** process. ATP (adenosine triphosphate) is the primary energy used for cellular metabolism.

Digestion is the process of dissolving food and breaking it down into simple chemical compounds in a living body, while metabolism is the process of building up the chemical changes in living cells by which energy is provided and assimilated into the system.

Through digestion, proteins are broken down into amino acids, carbohydrates to simple sugars, and fats into glycerol and free fatty acids (FFA) .

Several metabolic processes occur in the liver. Amino acids are removed from the blood and stored temporarily before transporting them through the body to synthesize new fish proteins. There is a continual breakdown and synthesis of proteins which are used in cell and tissue structures, and enzymes. Any excess will be used for maintenance/movement energy, broken down into carbon dioxide, water and ammonia. Absorbed sugars (from dietary carbohydrates) are converted to glucose in the liver, which can either be burned to supply energy or recombined to form glycogen stored in the liver as an energy reserve.

3.5.1. Functions of metabolism

Cellular metabolism has 4 specific functions.

1. Chemical energy, generally as ATP, can be obtained by the degradation of energy-rich nutrient biomolecules from the environment or by capturing solar energy.
2. To transform nutrient biomolecules into the building blocks or precursors needed for synthesizing macromolecular cell components.
3. To assemble such building blocks into proteins, nucleic acids, lipids and other cell components.

4. Forming and degrading biomolecules are required for the specialized functions of cells.

3.5.2. Phases of metabolism

The total metabolism of any cell can be broadly divided into 2 phases – catabolism and anabolism.

Catabolism refers to all the degradative reactions involving the degradation or breakdown of various complex molecules into smaller and simpler molecules. The chemical energy of metabolites is conserved in the form of ATP**, or catabolism, which is the process that results in the release of energy through the breakdown of nutrients, stored materials,** and cellular substances.

Anabolism refers to all synthetic reactions synthesizing various molecules from smaller and simple precursor molecules. This process requires the input of chemical energy in the form of ATP. Anabolism is the chemical process through which digested and absorbed products are used effectively to build or repair bodily tissues or restore vital substances broken down through metabolism.

The anabolic process uses monomers to build polymers. A **polymer** is a large, complex molecule made of similar small molecules. Those small molecules are called **monomers**. For example, amino acids, which are simple molecules (monomers) built through a series of anabolic chemical reactions, build **proteins**, which are large and complex molecules (polymers).

In living systems, both catabolism and anabolism occur concurrently and simultaneously. The energy released during catabolism is required for anabolic reactions and many other cellular activities like muscular contraction, active transport, osmotic works, etc. Catabolic reactions are oxidative, whereas anabolic reactions are reductive.

The pathway of reactions accessible to catabolism and anabolism is called the amphibolic pathway, which can be used catabolically.

3.5.3. Metabolism of energy in fish

Energy is needed for the maintenance of all living organisms. Most plants obtain their energy directly from the sun and use that energy to synthesize the complex molecules that make up the plant's structural and storage parts. Animals cannot utilize/radiant energy from the sun. They get the energy they need from the oxidation of the complex molecules that the animal eats. The energy in feed is not available until the complex molecules are broken down into simpler molecules by digestion. The products of digestion are then absorbed into the body of the animal, where oxidation processes occur, which release the energy.

Energy metabolism in fish is similar to that in mammals and birds, with two notable exceptions. These exceptions are:

a. Fish do not expand energy to maintain a body temperature different from their environment. Fish and shrimp, therefore, have much lower maintenance energy requirements than terrestrial farm animals.

b. The excretion of waste nitrogen requires less fish energy than in homeothermic land animals. Fish and shrimp can obtain 10–20% more energy from the catabolism of proteins than terrestrial farm animals, as they do not have to convert ammonia (the end product of protein catabolism) into less toxic substances (i.e., urea or uric acid) prior to excretion

There are considerable differences in the ability of different fish species to digest feed materials. Fish species range from strict herbivores, omnivores and carnivores. The food requirements of different species of fish vary greatly.

3.6. Energy flow in Fish

All food, whether natural or man-made, has a total or gross energy value. Not all the gross energy from nutrients is utilized, because some of it is not digested and will be lost as faecal energy material. Of the digestible components of the food, not all can be metabolized completely and further energy is lost through excretion in urine and through respiration across the gills, but these losses are relatively minor compared to the dietary energy excreted in the faeces. The remaining metabolizable energy can then be used for maintenance, movement and growth, although a small amount is lost through heat production.

Only a portion of the total chemical energy contained within the food ingested is available to the animal (i.e., net energy) for maintenance, activity and growth; 45% of the ingested food energy is lost as undigested food (faeces), metabolic excretion and heat. Although the energy losses will vary depending on the composition and digestibility of the feed ingredients used, feeding regime, water temperature, fish size and the physiological status of the animal, since all biological systems obey the laws of thermodynamics, the energy balance equation can be represented as follows:

Carnivores: $100C = 29P + 44R + 7U + 20F$

Herbivores: $100C = 20P + 37R + 2U + 41F$

Where C (for consumption) is the gross energy content of the food ingested, P is the energy utilized in growth materials (production), R is the net loss of energy as heat (for respiration), U (urinary loss) is the energy lost in nitrogenous excretory products, and F is the energy lost in the faeces.

The figures are expressed as a percentage of the energy consumed by ingested food.

Thus, approximately one-third of the diet is used by fish for growth. As muscle is essentially protein, more than 33% of the diet needs to be in the form of protein to ensure maximum conversion and growth. It should be remembered that energy needs for maintenance and voluntary activity must be satisfied before energy is available for growth. Providing the optimum energy level in diets for shrimp or fish is important because an excess or deficiency of useful energy can result in reduced growth rates. For example, excess dietary energy may result in decreased nutrient intake by the fish or excessive fat deposition in the fish. Also, during low food intake, fat and protein are withdrawn from the animal body to provide the energy needed for maintenance and the animal loses weight.

Not all of the gross energy from nutrients is utilized because some is not digested and absorbed for further metabolism. Thus, the amount of digestible energy (DE) a feed or feed ingredient provides is commonly expressed as a percentage of gross energy. A smaller fraction of the DE absorbed by the fish will be lost in metabolic wastes, including urinary and gill excretions, but these losses are relatively minor compared to the dietary energy excreted in the faeces. Because it is hard to collect fish urinary and gill excretions, it is much more challenging to determine metabolizable energy (ME) values for aquatic organisms than for terrestrial animals. Therefore, ME values are not commonly reported for fish feeds or ingredients.

Heat of nutrient metabolism, also called heat increment or specific dynamic action (SDA), is the increase in heat production following the consumption of food by an animal. It includes the energy expended in digestion, absorption, transportation and anabolic activities. It also includes the cost of excretion of waste products.

3.7. Specific Dynamic Action (SDA)

Digested food, particularly proteins, is not fully available to a fish even after it has been absorbed into the bloodstream. If amino acids are oxidized for energy, deamination (removal of the amino group) must occur first - a reaction requiring energy input. This process, known as specific dynamic action (SDA), can be measured externally in fish as an increase in oxygen consumption beginning soon after ingestion of food, followed by an increase in ammonia excretion.

The proportion of deaminated amino acids varies with the food and the fish's circumstances. Fish that are not growing because of low temperatures or have their ration at maintenance level or below would deaminate most or all of their amino acids. Fish kept at high rearing temperatures or high activity levels and, therefore, having very high metabolic rates would do the same. On the other hand, fish with rapid growth and high protein intake would deaminate a relatively small proportion of their digested protein, although the absolute quantity of deaminated amino acids could still be large enough to produce a relatively large SDA. The energy for deamination need not necessarily come from amino acids but will be preferentially taken from carbohydrates or lipids, if available. Thus, long ago, salmonid aquaculturists discovered this "protein-sparing" action of limited amounts of inexpensive carbohydrates in the diet to reduce the feed cost and still achieve the desired level of growth. The protein-sparing action of lipids appears to have been minimally investigated. One can thus minimize SDA costs but not avoid them altogether.

3.8. Metabolic pathways in fish

A metabolic pathway is a series of chemical reactions inside a living organism organized into metabolic pathways. These pathways result in the transformation of a particular molecule into a different molecule, or the synthesis (anabolism) and breakdown (catabolism) of most compounds occurs through several reaction steps. The reaction sequence is termed a *metabolic pathway*. Metabolic pathways include food digestion, energy storage, and other biochemical tasks.

Some pathways (e.g. glycolysis) are linear; others (e.g. the Krebs cycle) are cyclic. The changes at each step in a pathway are usually small and are promoted by efficient biological catalysts – the enzymes. In this way, the energy required or released at any given stage is minimal, which helps maintain a constant internal environment. Various feedback mechanisms govern metabolic rates.

3.8.1. Carbohydrates Metabolism

Carbohydrates are made up of carbon, hydrogen and oxygen atoms and are classified as mono-, di- and polysaccharides, depending on the number of sugar units they contain. Following digestion, carbohydrates are absorbed through the wall of the digestive tract. They are absorbed and enter the bloodstream as monosaccharides (glucose, galactose and fructose). In vertebrates, the absorbed nutrients are transported to the liver, where the initial metabolism processes occur. Glucose is the most important of the monosaccharides, acting as a major metabolic energy source that circulates in all vertebrates. They may be utilized directly for energy by all tissues, temporarily stored as glycogen in the liver or muscle, or converted to fat, amino acids and other biological compounds. Some tissues (brain) use only glucose as an energy source, so maintaining blood glucose levels is essential. Glucose is the primary energy source for the brain, skeletal muscle and red blood cells. Deficiency can impair the brain and nervous system.

Four reaction pathways exist to do this: Carbohydrate metabolism begins with glycolysis, which releases energy from the breakdown of glucose or glycogen from carbohydrates to form two molecules of pyruvate, which enter the Krebs cycle (citric acid cycle), an oxygen-requiring process, through which they are completely oxidized. Glycolysis occurs in the cytosol (fluid portion) of a cell and can proceed regardless of the presence of oxygen. In the first glycolysis stage, energy is used to phosphorylate the 6-carbon glucose molecule. This means that phosphate is taken from ATP (which becomes ADP) and added to the glucose molecule.

Before the Krebs cycle can begin, pyruvate loses a carbon dioxide group to form acetyl coenzyme A (acetyl-CoA). This reaction is irreversible and has critical metabolic consequences. The conversion of pyruvate to acetyl-CoA requires the B vitamins and then proceeds through the Krebs cycle to produce ATP, CO_2 and water. The hydrogen in carbohydrates is carried to the electron transport chain, where the energy is conserved in ATP molecules. The metabolism of one molecule of glucose yields thirty-one molecules of ATP. The energy released from ATP through hydrolysis (a chemical reaction with water) can be used for biological work. Only a few cells, such as liver and kidney cells, can produce their glucose from amino acids, and only liver and muscle cells store glucose in the form of glycogen. Other body cells must obtain glucose from the bloodstream. Glycolysis has a dual role. It degrades monosaccharides to generate energy and provides **glycerol** for **triglyceride** synthesis. The Krebs cycle and the electron transport chain occur in the **mitochondria.** The Krebs cycle and the electron transport system produce most energy derived from carbohydrates, protein and fat. Krebs cycle, or the citric acid cycle, is part of a metabolic pathway involved in the chemical conversion of carbohydrates, fats and proteins into carbon dioxide water to generate usable energy.

3.8.2. The Electron Transport Chain

Very little energy has been produced during glycolysis and the Krebs cycle. The electron transport chain and oxidative phosphorylation will release most energy locked in the original glucose molecule. The electron transport chain is a network of electron-carrying proteins located in the inner membrane of the mitochondrion. These proteins transfer electrons from one to another, down the chain, much like a bucket brigade passes buckets of water. These electrons, along with protons, will eventually be added to oxygen, the final electron acceptor. This produces water but does not produce any ATP. A proton motive force produces the ATP. This force stores potential energy created by the gradient formed when hydrogen (protons) is moved across a biological membrane. Therefore, the electron transport chain merely produces a gradient through which ATP can be made (this is known as chemiosmosis). The electron transport chain produces the remaining 32-34 ATP.

Under **anaerobic** conditions, glycolysis produces lactate from pyruvate in active muscle. When energy demands exceed the oxygen supply, this reaction is essential in the muscle. The lactate moves through the bloodstream to the liver, which is oxidized back to pyruvate. Glyconeogenesis converts pyruvate to glucose, which is carried back to the muscles. The Cori cycle is the flow of lactate and glucose between the muscles and the liver.

Glycogenesis is the conversion of excess glucose to glycogen. Glycogenolysis is the conversion of glycogen to glucose, which could occur several hours after a meal, overnight in the liver, or the absence of glucose-6-phosphate in the muscle to lactate. Glyconeogenesis is the formation of glucose from non-carbohydrate sources, such as certain amino acids and the glycerol fraction of fats, when carbohydrate intake is limited. The liver is the main site for glyconeogenesis, except during starvation, when the kidney becomes vital in the process. Disorders of carbohydrate metabolism include diabetes mellitus, lactose intolerance and **galactosemia.**

3.8.3. Protein Metabolism

Proteins contain carbon, hydrogen, oxygen, **nitrogen** and sometimes other atoms arranged as strands of amino acids. They form the cellular structural elements as biochemical catalysts and important regulators of **gene expression.** Nitrogen is essential to forming twenty different amino acids, the building blocks of all body cells. Amino acids are characterized by a terminal carboxyl group and an amino group in the alpha position, and peptide bonds connect them.

Protein anabolism is the process of forming proteins from amino acids (anabolic amino acid synthesis). Protein catabolism, or proteolysis, is the process by which proteins are broken down into amino acids.

Digestion breaks protein down to amino acids. If amino acids are in excess of the body's biological requirements, they are metabolized to glycogen or fat and subsequently used for energy metabolism. If amino acids are to be used for energy, their carbon skeletons (A carbon skeleton is a linkage or chain of carbon atoms) are converted to acetyl coenzyme A (acetyl CoA), after deamination, which enters the Krebs cycle for oxidation producing ATP. The final products of protein catabolism include carbon dioxide, water, ATP, urea and ammonia.

3.8.4. Fat (Lipid) Metabolism

Fats contain mainly carbon and hydrogen, some oxygen and sometimes other atoms. Dietary lipids are a complex mixture of the neutral lipids - glycerides (principally triacylglycerol [triglyceride], the form in which fat is stored for fuel), **phospholipids, sterols** (principally **cholesterol**) and other complex lipids. Fats provide 9 kilocalories per gram (kcal/g), compared with 4 kcal/g for carbohydrate and protein. Triacylglycerol, whether in the form of chylomicrons (microscopic lipid particles) or other **lipoproteins,** is not taken up directly by any tissue but must be hydrolyzed outside the cell to **fatty acids** and glycerol, which can then enter the cell.

Fatty acids come from the diet, adipocytes (fat cells), carbohydrates and some amino acids. After digestion, most fats are carried in the blood as chylomicrons. The main pathways of lipid metabolism are lipolysis, beta-oxidation, **ketosis** and lipogenesis. Lipolysis (fat breakdown) and beta-oxidation occur in the mitochondria. It is a cyclical process in which two carbons are removed from the fatty acid per cycle as acetyl CoA, which proceeds through the Krebs cycle to produce ATP, CO_2 and water. Ketosis occurs when the liver's formation rate of ketones is greater than the tissues' ability to oxidize them. It occurs during prolonged starvation and when large amounts of fat are eaten without carbohydrates. Lipogenesis occurs in the cytosol. The main sites of triglyceride synthesis are the liver, adipose tissue and intestinal mucosa. The fatty acids are derived from the hydrolysis of fats and the synthesis of acetyl CoA through the oxidation of fats, glucose and some amino acids. Lipogenesis from acetyl CoA also occurs in steps of two carbon atoms. NADPH produced by the pentose-phosphate shunt is required for this process. Phospholipids form the interior and exterior cell membranes essential for cell regulatory signals.

A stage in fat metabolism is stimulated primarily when the liver is low on glycogen (stored glucose). The breakdown of fats in this process results in three different molecules called "ketones" or "ketone bodies", two of which (acetoacetate and ß-hydroxybutyrate) can be used as fuel by most tissues of the body. Some tissues, such as the heart muscle, prefer ketones to glucose. The third ketone, acetone, cannot be used by the body and is excreted in the urine and breath. In organic chemistry, a **ketone** is a compound with the structure RC(=O)R′, where R and R′ can be a variety of atoms and groups of atoms. It features a carbonyl group (C=O) bonded to two other carbon atoms. Acetone is the simplest example of a ketone, and the word *ketone* derives its name from *Aketon*, an old German word for acetone.

Ketones differ from aldehydes because the carbonyl is placed between two carbons rather than at the end of a carbon skeleton. They are also distinct from other functional groups, such as carboxylic acids, esters and amides, with a carbonyl group bonded to a heteroatom.

Fats are readily catabolized for daily maintenance and movement energy needs. The fat deposition will occur if there is too much fat in the diet or fish are overfed for a long time. Fats can be synthesized from amino acids, glucose, glycerols, and free fatty acids - it is thus easy for fish to accumulate fat from their diet. Fat is also the most economical way to store energy as it has inherent higher energy latency.

3.9. Role of enzymes in metabolic pathway

Enzymes play an essential role in the various metabolic pathways. They are present as coenzymes with some other group attached to them having substrate specificity to accelerate the particular metabolic reaction. Apart from that, various co-factors also played an essential role in enhancing their mechanism of action. Specific enzymes, like the digestion of food, carry out particular types of metabolic reactions. In that way, food is comprised of different components. For example, carbohydrates are acted upon by enzymes like maltase, sucrase, galactase, etc., to break down into simple sugar.

Enzymes also play an essential role in DNA and RNA synthesis like ribonucleotide reductase to convert ribonucleic acid into deoxyribonucleic acid and reverse transcriptase to form cDNA from mRNA

3.10. Metabolic rate

It is defined as the measure of the energy used by an animal in a given time. It is also defined as the conversion of chemical energy into heat or heat energy released per unit of time.

Three levels of metabolism will be distinguished here. Following the current usage among several fisheries workers, these will be termed "basal ", "standard", and "active" levels of metabolism.

3.10.1. The basal metabolism or Basal Metabolic Rate

The basal metabolism or Basal Metabolic Rate is the minimum rate of energy required by the body to keep the organism alive or in other words to sustain basic life processes, including breathing, circulation and tissue repair the minimum rate of energy expenditure to keep the organism alive expressed as mg of O_2/kg/hr.

3.10.2. Standard metabolism or Standard Metabolic Rate (SMR)

Standard metabolism or Standard Metabolic Rate (SMR) refers to an animal's resting and fasting metabolism at a given body temperature minimum rate of oxygen consumption of an organism at rest in the post-absorptive state (when there is no food in the gut) expressed as mg of O_2/kg/hr. Standard metabolism is an approximation of the minimum rate for the intact organism.

3.10.3. Active or field metabolism

Active or field metabolism is the maximum rate of energy expenditure at maximum sustained activity rate of oxygen consumption during maximum sustained activity expressed as mg of O_2/kg/hr. It is the average energy utilisation rate as the animal goes about its regular activities, which may range from complete inactivity while the animal is at rest to exerting maximum energy.

Routine metabolism is the metabolic rate of an organism during regular spontaneous activity expressed as mg of O_2/kg/hr.

Most measures of metabolism have been measurements of routine metabolism. Standard and active metabolism has not yet been measured, and these limits have been less often well worked out.

3.11. Measuring the metabolic rate of an organism

Measuring metabolic rate is helpful to physiologists and ecologists, animal behaviourists, evolutionary biologists and many other fields of study because the metabolic rate can yield information on the energy requirements of animals. In order to learn more about the heat-conserving and heat-releasing mechanism of animals, metabolic rates can be measured at different temperatures. Metabolic rate can tell us how much energy an animal needs to expend to fly, swim, run or walk. It can also tell us something about the processes of tissue growth and repair, as well as the chemical, osmotic, electrical, and mechanical internal and external work necessary for such things as locomotion and communication.

The metabolic rate of an animal (oxygen consumption) is primarily influenced by water temperature, fish size and activity of the fish, feeding, osmoregulation and concentration of oxygen in water. The concentration of ammonia, nitrite, or other adverse compounds in the water, nutritional deficiencies, hormone applications, and other factors can also affect fish metabolism rates.

Among all the environmental factors that affect the physiology of animals, temperature is most important because with every 10°C rise in temperature, metabolism doubles.

UNIT IV

CHAPTER 1

REPRODUCTIVE PHYSIOLOGY

1.1. Reproduction

Reproduction is the biological process of producing new "offsprings" or individual organisms from their **"parents".** Reproduction is a fundamental feature of all known life; each organism exists as the result of reproduction.

1.2. Importance of reproductive physiological studies

1. To develop maturation under captive condition

Most of the females cannot mature under captive conditions. Generally, the organism should be bred in captivity to follow good farming practices.

2. To enhance sperm count and viability

Although male fish mature under captive conditions, their sperm count is reduced. The sperm viability enhancement will result in better hatching.

3. Control of excess and undesirable spawning

Fishes, like tilapia, can spawn in rearing units. Because of this, the growth will be minimal after maturation as most of the energy is directed towards gonadal growth. For better biomass production, and excessive spawning must be controlled. The production of sterile fish is the way to control undesirable spawning.

4. Seasonality of spawning

Fishes spawn only during a particular season. Indian Major Carps (IMC) breed only during the monsoon seasons. The culture practice is a year-round operation. So, the seeds must be available throughout the year. A continuous supply of matured fish is required to get year-round production.

5. Synchronous spawning

In captive conditions, male and female fish will not spawn simultaneously, i.e., once the male responds, the female will not spawn, and vice versa. This condition affects the spawn quantity and quality. Both males and females have to respond simultaneously for better spawning. Several studies were required to help induce synchronous spawning in fishes to obtain a better production.

6. The seed quality improvement

In nature, the availability of seeds is poor. The collected natural seeds are not uniform in size, have a mixing of genetic characteristics and may carry harmful pathogens. To overcome this problem, a quality seed can be produced under hatchery conditions for which reproductive physiological study will significantly help.

1.3. Types of reproduction

3 types of reproduction are possible:

1. Bisexual reproduction
2. Hermaphroditic reproduction
3. Parthenogenetic reproduction

1.3.1. Bisexual reproduction

Fishes are generally bisexual, and sperms and eggs develop in separate individuals, which have either testes or ovaries. E.g., Carps.

1.3.2. Hermaphroditic reproduction

The eggs and the sperms develop in the same gonad, and self-fertilization occurs. There are 2 types of hermaphroditism.

1.3.2.1. Synchronous or simultaneous hermaphroditism

Synchronous or simultaneous hermaphroditism is the condition in which both parts of the ovotestis mature simultaneously. This gives rise to a fish which is functionally both male and female. (e.g., Coral reef fish). Fish of this species usually form pairs with a single partner. Each individual spawns alternatively as male and female. Shortly before spawning, the fish ready to assume the male role gives a courtship call, and the fish assuming the female role responds and releases eggs. Following fertilization of the eggs, the spawning pair breaks apart and takes cover on the reef. In the next spawning event, the roles of the 2 fish are reversed, with the original male acting as the female and the original female taking over the role of the male.

1.3.2.2. Successive hermaphroditism

Sequential or successive hermaphrodites are born with one sex and change sex sometime during life. Successive hermaphroditism in which the fish undergoes sex reversal.

Protandrous- males change their sex to become females.

Protogynous- females change their sex to become males.

a. Protandrous

An example of protandry is found in the anemone fishes. The fishes live with anemones in a symbiotic relationship; the anemone provides the fish with shelter and protection from predation, and the fish supply the anemone with food. Groups of fishes will live with one anemone and will not switch anemones. Only the two largest will mate: the largest female and the second largest male. With the female being the largest, she can produce the most eggs. When the female dies, the largest male will change sexes and become the female. The rest of the fish are immature males. Some sea basses (*Serranidae*) are protandric hermaphrodites, being males at first and females later. E.g., *Sparidae* and *Serranidae* are true hermaphrodites.

b. Protogynous

A classic example of protogyny is found in the wrasses and parrotfishes. The males in these species form harems, with one large male sequestering and defending a group of smaller females. The male enjoys spectacular reproductive success, as it has many females to mate with. The females also enjoy limited reproductive success, producing as many eggs as possible, all fertilized by one male. The male has the advantage over the females; many females produce eggs for him to fertilize, whereas the females only have themselves.

For example, some damselfishes (family - Pomacentridae) begin life as males and change into females. In some, females can revert back to males.Some seabasses (family Serranidae) change from female to male and can be reverted to female. Most wrasses (family - Labridae) are born female, grow into sexually mature females, and have the potential to transform into functional males later in life. In many of the wrasses, sex change correlates with social hierarchy and social behaviour: social structure includes a large dominant male and many smaller, subordinate females. Removing the male from the group triggers the largest female to transform into a male.

1.3.3. Parthenogenetic reproduction

Parthenogenesis is the development of young without fertilization. This is also called gynogenesis. Males of other species are required for sexual reproduction, whereas the sperms only stimulate gynogenetic development without contributing their chromosomes (it does not take part in heredity). Therefore, the young are always female and have no trace of paternal characteristics. Eg. *Poecilia formosa* (Guppies).

Some species are unisexual. In unisexual species, there is no fusion of sperm and egg. The sperm is necessary to trigger the egg cell to develop into an adult, but the sperm cell ultimately degenerates and does not take any part in heredity. The resulting youngones are always females with no trace of paternal characteristics. Thus, unisexual species are entirely female. They mate with males of related species to produce female offspring. *Poecilia formosa* is an example of a unisexual species. Always female, *P. formosa* mates with male *P. mexicana* or *P. latipinna.*

1.4. Male reproductive system

The male fish's reproductive organ consists of a pair of testes. The cartilaginous fishes are usually elongated ovoid structures located anteriorly and attached to the body cavity by means of a mesorchium.

In most bony fish, the testes are whitish or creamy-coloured, elongated lobulate organs attached to the dorsal body wall by mesochoria. A vas deferens or sperm duct arises from the posterior region of each elongated testis and leads to the urinary papilla which is located between the rectum and the urinary ducts.

In most cases, testes are a pair of elongated structures composed of numerous seminiferous tubules embedded in the stroma. The testis consists of thin-walled tubules or lobules that contain germ cells - the spermatogonia - which are endodermal in origin. Germ cells divide in clusters enclosed by a cyst. Primary spermatogonia - present throughout the year, divides mitotically to give rise to secondary spermatogonia, which is transformed into primary spermatocytes. They divide by meiosis and give rise to spermatids from which spermatozoa are formed. The seminiferous tubules are packed with spermatozoa in the pre-spawning and spawning periods.

Structure of sperm

a) Headpiece b) Middle piece c) Tail

In fish sperm, acrosomes are absent.

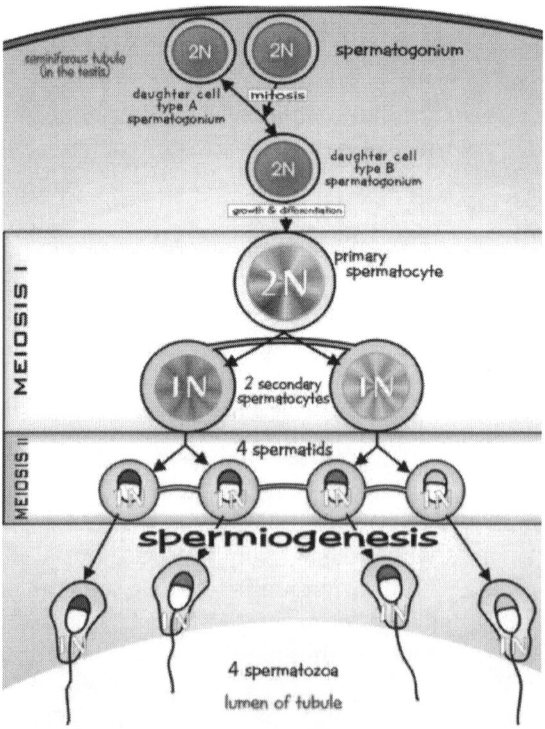

Fig 4.1 - Spermiogenesis

1.4.1. Reproductive cycle (Testicular)

Most fish breed in one season yearly (seasonal spawners), and some breed throughout the year (year-round spawners). Seasonal breeders exhibit rhythmic changes in the structure and physiology of the ovary and testes in different seasons. These changes in testis are demarcated into five phases

1. **Resting phase:** Testis remain in an immature state. Seminiferous tubules are solid being filled with spermatogonial cells
2. **Preparatory phase:** During this phase, cell proliferation and 1st and 2nd meiotic division occur. As a result, primary spermatocytes, secondary spermatocytes and spermatids are produced (spermatogenesis).
3. **Mature Phase:** During this phase, spermatids undergo further development and metamorphosis and develop into mature spermatozoa or sperm. The process of development of spermatozoa from spermatids is called spermiogenesis.
4. **Spermiation Phase:** During the courtship and mating process, the male ejects milt (spermatozoa in seminal fluid) out of its body through the genital aperture to fertilize the eggs the female releases. The seminal fluid is secreted mainly by the cells lining the vas deferens (sperm duct) and nourishes the mature sperm. Both spermatogenesis and seminal fluid secretion are controlled by gonadotropin of the pituitary gland and male hormone testosterone.
5. **Post-spermiation phase:** During this phase, the testis is characterized by evacuated seminiferous tubules.

1.5. Female reproductive organ

Female reproductive organs are composed of

1. Ovaries: paired, an elongated sac-like structure on either side of the ventral to the kidney in the abdominal cavity and found attached to the body wall through a mesovarium
2. Oviduct
3. Genital pore or cloaca

1.5.1. Types of ovaries

1. **Cystoarian:** Mature oocytes released in the ovoceal that is continued as oviduct to open into the exterior (Lepidosteus and Teleost fishes)
2. **Gymnoarian:** Mature oocytes released into the body or coelomic cavity from where they pass out through the genital aperture (Chondrichthyes, Dipnoi, Chondrostei and Amia)

The ovary in most teleost fishes is a hollow sac-like organ that extends numerous ovigerous folds (lamellae) lined by germinal epithelium. The germ cells, oogonia (originate from germinal epithelium), multiply mitotically and transform into nonyolky primary oocytes. Primary oocytes, covered generally by two layers of follicle cells, an outer thecal layer and an inner granulosa layer, undergo vitellogenesis when the yolk is deposited in the ooplasm. Analysis of yolky eggs indicates that the egg yolk is composed of lipovitellin and phosvitin; the former is rich in lipid and poor in phosphorus, and the latter is rich in phosphorus and poor in lipid.

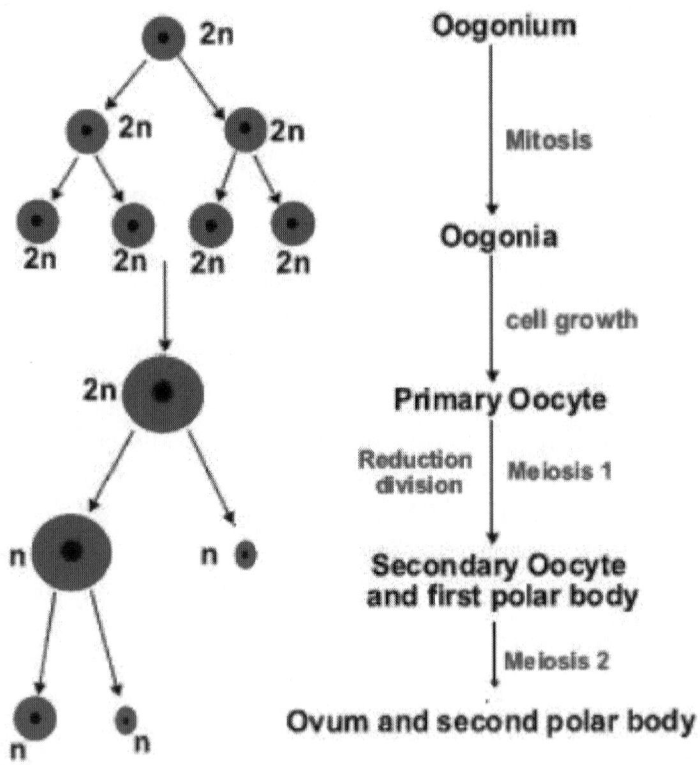

Fig 4.2 - Oogenesis

Parallel to the testicular cycle, the primary reproductive organ of the female, i.e., the ovary, also undergoes rhythmic change, which is divided into five phases as follows:

1. **Resting Phase:** The ovary is immature, containing nests of oogonial cells, stage 1 oocytes at different growth phases, and a few stage II oocytes. During this stage, the first meiotic division of the nucleus is initiated and arrested at the pachytene stage.
2. **Preparatory phase:** Oocytes are in stage III. There is a beginning of vitellogenesis in the oocyte cytoplasm. Oocytes get differentiated with the thecal and granulosa layers, assuming the steroidogenic function. Cells of the granulosa layer start producing the female hormone, estradiol.

3. **Maturing or Pre-spawning phase:** Characterized by the intensive process of vitellogenesis by which the ooplasm of an oocyte is loaded with yolk granules and in stage IV. 1st meiotic division is completed, and the second meiotic division occurs. The size of the ovary and GSI will be maximum at this phase.
4. **Spawning Phase:** Characterized by gravid ovary containing ripe oocytes. During spawning, the follicles of fully ripe oocytes rupture. As a result, oocytes are released into the ovocoel (cystovarian ovary) or body cavity (gymnovarian ovary), from where they pass out through the genital pore into water.
5. **Post-spawning phase:** The ovary exhibits a collapsed appearance as evacuated follicles are seen after the release of eggs.

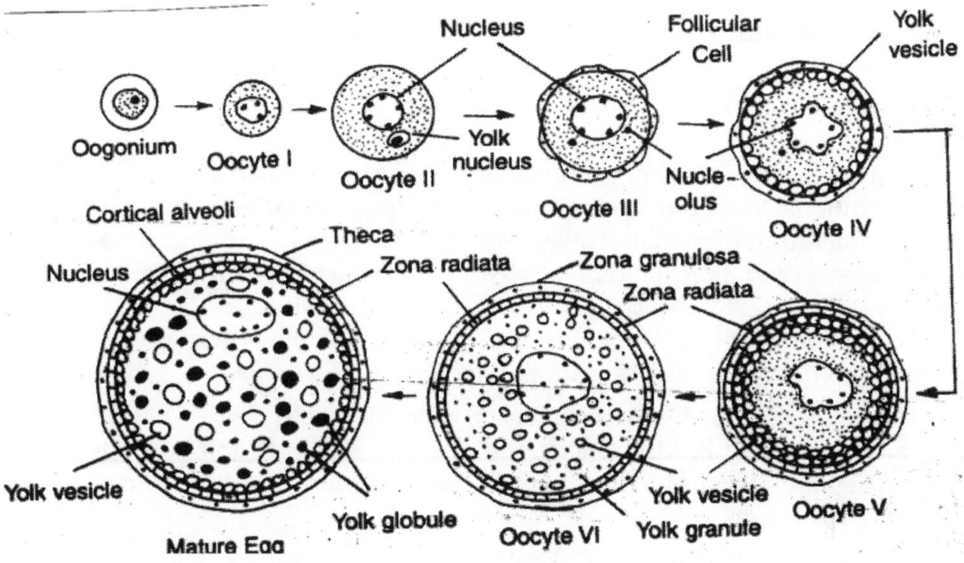

Fig 4.3 - Stages in the maturation of oocytes

1.5.2. Vitellogenesis

Accumulation of yolk substances in eggs is called vitellogenesis. At the beginning, there is no yolk. The blood of mature females contains a calcium-binding lipophospo-protein called vitellogenin, which is synthesized in the liver by stimulating female sex steroids. It passes from the plasma to the oocyte and undergoes limited proteolysis, giving rise to yolk protein lipovitellin (Lv), phosvitin (Pv), '-component ('-c), and C-terminal peptide. Yolk protein is deposited in three yolk substances: Yolk vesicles containing glycoprotein, yolk globules-lipoprotein, and oil droplets- Glycerides. After completion of vitellogenesis, movement of germinal vesicles to animal pole, Fusion of yolk granules and grouping of oil droplets occurs.

1.5.3. Gonadosomatic Index (GSI)

The following formula can calculate the Gonadosomatic index of a species

GSI= (Weight of the gonads/weight of the fish) X 100

The gonadosomatic index of a species has been widely used to indicate the maturity and periodicity of the spawning of the fish. The GSI increases with the maturation of the fish and is maximum during the peak period of maturity. It decreases abruptly after spawning.

1.6. Sex determination

The reproductive systems and patterns of sexual behaviour fish display are highly diverse. There are also several different genetic systems and sex-determination mechanisms.

As with other animals, the genetic material of the fish is located within the cell nucleus in the form of chromosomes. The number of chromosomes varies from species to species, but the chromosome number almost invariably remains constant within a species. In most fish, the chromosomes occur in pairs- diploid (2N) condition. One of the chromosomes of each pair derives from the male parent and the other from the female. Nevertheless, there are some exceptions. Some fish, e.g., salmonids and catastomids, may have evolved via tetraploidy, which involved doubling the original diploid number (2N) of chromosomes to the tetraploid (4N) condition. Some fish have become triploids, but the occurrence of natural populations of triploid fish appears extremely rare.

Our bodies are made up of millions of cells. Most cells contain a complete set of genes. Genes act like instructions, controlling our growth and how our bodies work. They are responsible for many of our characteristics, such as eye colour, blood type or height.

Genes are carried on thread-like structures called chromosomes. Usually, we have 46 chromosomes in most cells. We inherit our chromosomes from our parents, 23 from our mother and 23 from our father, so we have two sets of 23 chromosomes or 23 pairs. Because the chromosomes comprise genes, we inherit two copies of most genes, one from each parent. This is why we often have characteristics similar to those of our parents. The chromosomes, and therefore the genes, comprise a chemical substance called DNA.

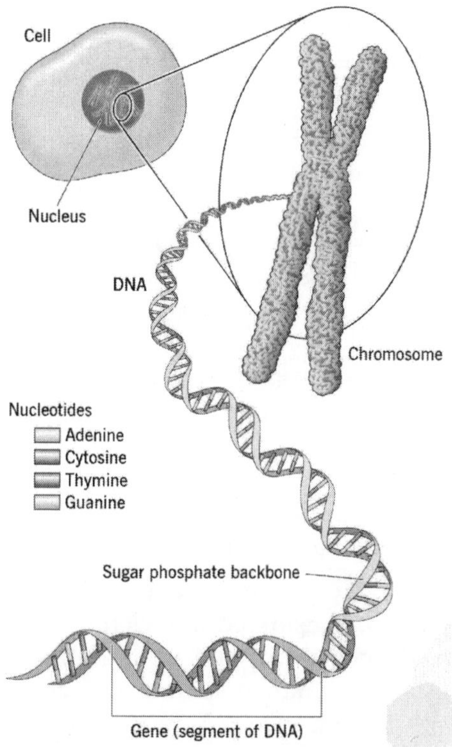

Fig 4.4 - Genes, chromosomes and DNA

The chromosomes numbered 1 to 22 look the same in males and females. These are called the autosomes. Pair number 23 is different in males and females, and they are called the sex chromosomes. There are two kinds of sex chromosomes, one called the X chromosome and the other called the Y chromosome. Females typically have two X chromosomes (XX). A female inherits one X chromosome from her mother and one X chromosome from her father. Males typically have an X and a Y chromosome (XY). A male inherits an X chromosome from his mother and a Y chromosome from his father.

A **chromosome** is an organized structure of DNA and protein found in cells. It is a single piece of coiled DNA containing many genes, regulatory elements and other nucleotide sequences. Chromosomes also contain DNA-bound proteins, which package the DNA and control its function.

Chromosomes are long, stringy aggregates of genes that carry heredity information. They are composed of DNA and proteins and are located within the nucleus of our cells. Chromosomes determine everything from hair colour and eye colour to sex. Whether you are a male or female depends on the presence or absence of specific chromosomes.

1.6.1. Chromosomes in fish

Goldfish have 94 while carp have 104, a crayfish 200 - Chromosome numbers vary from species to species

The XY sex chromosomes are different in shape and size from each other, unlike the autosomes, and are termed allosomes.

1.6.2. Sex chromosomes

The most common system of sex determination in fish, as in other vertebrates, is known as the XY system. The homogametic individuals (XX) are female, while heterogametic (XY) individuals are males.

In the second system, the male is homogametic (ZZ), and the female is heterogametic (WZ). This system is known as the WZ system.

1.6.3. Autosomes

The sex chromosomes contain genes that determine the gender of an individual. The remaining chromosomes that are not involved in determining the sex of an individual are called autosomes.

Any chromosome is not considered a sex chromosome or is not involved in sex determination. It occurs in pairs in somatic cells and singly in sex cells (gametes). For instance, a human somatic cell generally contains 23 pairs of chromosomes (total = 46 chromosomes). Twenty-two (22) pairs will be autosomes, and only one will be a pair of sex chromosomes (X and Y).

A light press on the belly of most ripe fishes will bring the whitish milt or the eggs into view near the anus. However, fish also use other features to exhibit the sexes.

The characteristics of sexual difference or sexual dimorphism that enable the identification of the sexes are classed as primary and secondary. Primary sexual characters are those that are concerned actually with the reproductive process. Testes and their ducts in the male and ovaries and their ducts in the female constitute primary sexual characters. The primary sexual characters often require dissection for their discernment, which makes the secondary sexual characters often more helpful.

1.6.4. Outstanding differences

- Body shape is an important secondary character. The female is pot-bellied compared to the male, mainly when she is ripe or near ripe for spawning because of her ovaries.
- Many fish develop a "nuptial dress," which is particularly clearly expressed in the salmonids. During the reproductive period, the change is first expressed in the colouration, which becomes much brighter and in which colours appear that were not present. Particularly in the males, the skeleton

of the jaws becomes bent and large teeth appear on them. In the cyprinoids, the nuptial dress is expressed in the appearance of a brighter colouration and of peculiar tubercles on the horny epithelium of the head and scales of the body. Very significant changes occur in the colouration of many gobies.

- The paired and the unpaired fins are slightly larger in males than in females.
- In some fishes, the lower lobe of the caudal fin is greatly extended. E.g., Swordtail.
- In the males of several species, the anal fin becomes enlarged into an intromittant, copulatory organ designated as gonopodium. E.g., Guppy, *Gambusia affinis* (Mosquito fish).
- The pelvic fins of sharks are modified as claspers or myxopterygia.
- Colouration also serves sexual distinction, which is termed sexual dichromatism. In general, males are brighter in colour than females.
- Head characteristics also serve to distinguish the sexes among fishes. In the chimaeras, the male develops a spiny stout, retractile knob, the frontal clasper on the upper part of the head, which has similar characteristics to forehead brooders.
- In deep-sea angler fish, the male is minute and is parasitic, and attached to a larger female

1.7. Maturation and spawning

Testes and ovaries mature after undergoing well-marked histo-morphological alterations and increase in size. This is called maturation, and the final expulsion of sperm or ova from the body into the surrounding water is called "spawning", followed by fertilization. Each species spawns during a specified period under favourable conditions. The part of the year during which the gonads attain full maturity, and spawning occurs in the population is called the species' breeding season. After spawning, the residual germ cells rest briefly and enter the maturation cycle again to produce gametes for the next season.

Teleosts vary in the rhythm of the maturation of oocytes. Some species spawn only once, others twice, while others may spawn several times a year. Many Indian marine fishes appear to be continuous breeders and have a protracted spawning period of 7-9 months. Maturation and spawning periods vary in different species and even in the same species collected from different places. Temperature, salinity, photoperiod and duration are considered important factors controlling the maturation of gonads. Most of the Indian freshwater teleosts spawn after the onset of monsoon. The Asian seabass, *Lates calcarifer,* spawns from June to October, while the grey mullet, *Mugil cephalus,* spawns during monsoon seasons from October to June.

1.8. Ovulation and fertilization

The mature oocyte projects into the lumen of the ovary. Then, the follicle ruptures, and the egg escapes from the aperture. This is called **ovulation**, which refers to expulsing the mature oocyte from its follicle into the ovarian cavity. The forces causing the rupture of the follicle are not known, and the muscle fibres of the ovarian wall may be responsible for it. Fish ovaries may be of three types: gymnovarian, secondary gymnovarian and cystovarian. In the first type, the oocytes are released directly into the coelomic cavity, which then travel through the ostium, enter the oviduct, and are eliminated. In the secondary gymnovarian ovaries, ova are shed into the coelom and then go directly into the oviduct. In the third type, the oocytes are directly deposited into the exterior by travelling through the oviduct. A female fish may have one of these three types of ovaries, depending on the species.

Gymnovaries are the primitive condition found in lungfishes, sturgeons, and bowfins. Cystovaries are the condition that characterizes most of the teleosts, where the ovary lumen has continuity with the oviduct. Secondary gymnovaries are found in salmonids and a few other teleosts.

Ostium means an opening derived from the Latin word for a door or opening.

Fertilization is usually external in water, and sperm and ova are discharged close to each other. Sperms become very active soon after being released in water, but survive only for a few minutes during fertilization.

The released egg is protected by a reasonably challenging chorion or egg case. Within this, the cytoplasm and yolk are contained by a vitelline membrane. Often, one or more oil globules are present. The chorion has only one opening, the micropyle, for the entrance of sperm. Usually, the **micropyle** closes after the entry of one sperm to prevent multiple fertilization. They are collectively referred to as egg activation. The vitelline membrane separates from the chorion, creating a perivitelline space. Some elasmobranchs have polyspermy in which several spermatozoa may enter the micropyle, but only one fuse with the egg's nucleus, and the rest are being reabsorbed and perhaps used as an additional nutrient. After fertilization the chorion hardens. The egg goes through cleavage and morphogenesis as the cells divide and form layers, followed by organs.

1.9. Embryonic Development in Fish

Fish embryonic development consists of seven stages leading to hatching. These stages are the **zygote period, cleavage period, blastula period, gastrula period, segmentation period, pharyngula period, and finally hatching**.

Zygote period: Soon after fertilization, the micropyle is closed, thereby preventing the entry of any other sperm into the oocyte. The perivitelline space gets filled with water and colloid, and oocyte swelling occurs. This phenomenon is called water hardening. During water hardening, the animal pole develops a small hillock of protoplasm on the yolk mass known as egg blastodisc.

Cleavage period: Cleavage furrow is limited to the cytoplasm of animals, and this type of cleavage is called meroblastic cleavage. As successive cleavage continues, a group of cells is formed, and this group is called morula.

Blastula period: After further division, the cell is arranged into a blastoderm layer. Each cell of the blastoderm is called a blastomere. Cleavage also forms another type of cell called periblast, which lies between the blastoderm and yolk. The embryo develops from the blastoderm. A space between the blastoderm and periblast is called a segmental cavity or blastocoel. At this stage, the embryo is called a blastula. Presumptive germ layers (ectoderm, endoderm and mesoderm) can be noted at the blastula stage.

Gastrula period: The process of formation of gastrula is called gastrulation. At this time, the under rim of blastodisc thickens to form a marginal ridge and the inner layer of this is called the germ ring. The germ ring, called the embryonic shield, is the thickest at the caudal end. Gastrulation occurs by three processes **invagination, involution and epiboly.** An invagination appears at the future endodermal region of the blastula. The portion that is invaginated grows below the blastoderm, and this portion is called hypoblast. The endoderm and mesoderm cells multiply and migrate inward over the lip of the blastopore, which is called involution. During this process, the blastoderm cells also continue to grow over the yolk, which is called epiboly. As a result, the yolk mass is covered by a layer of cells called epiblasts, and future ectodermal cells are involved in this process. The periblast also grows over the yolk, forming the inner layer covering the yolk below the epiblast. The inner periblast and outer epiblast layer enclose the yolk sac. With the formation of the yolk sac, gastrulation is completed.

Segmentation period: The primary organ rudiments appear in the anterior part of the embryo. The anterior part of the neural plate from where the brain develops appears first. Notochord gets separated from mesoderm. Mesoderms lie on both sides of the notochord and are segmented, forming various body organs. This results in a small embryo with a cylindrical and bilaterally symmetrical body.

Pharyngula period: In this stage, the embryo exhibits fully formed organs. Ectoderm gives rise to the epidermis and its derivatives like the enamel of teeth, olfactory epithelium, lens of the eye, inner ear, brain, spinal cord, retina and optic nerve. The mesoderm is divided into the dorsal epimere, middle mesomere, and ventral hypomere. Epimere becomes divided into somites, each having three parts: outer dermatome, middle myotome and inner sclerotome. Dermatomes give rise to connective tissue and muscles of the dermis of the skin. Myotome gives rise to the muscle of the trunk, appendicular skeleton, fins and their muscles. The vertebral column is derived from the sclerotome. Kidney, gonads and their ducts are derivatives of mesomere. Hypomere develops into somatic and splanchnic layers of mesoderm and encloses a coelomic cavity. Mesenchyme cells derived from the splanchnic layer develop into involuntary muscles and connective tissue of the gut, head and blood vessels. Mesenchyme of the head region gives rise to the skeleton and muscle of the head and the eye's outer layer. At the end of this stage, the embryo becomes active and exhibits continuous twitching movement, showing that embryonic development is completed.

Hatching: Hatching involves two mechanisms: 1) mechanical and 2) enzymatic. In mechanical hatching, egg envelopes are broken down by mechanical action such as pressure excreted from within or mastication by the embryo itself. In enzymatic hatching, young emergence occurs after the egg envelope's dissolution or softening by the enzyme secreted by the embryo. This enzyme is called the hatching enzyme. These enzymes are secreted into perivitelline fluids from hatching gland cells in the embryo's epidermis at the time of hatching.

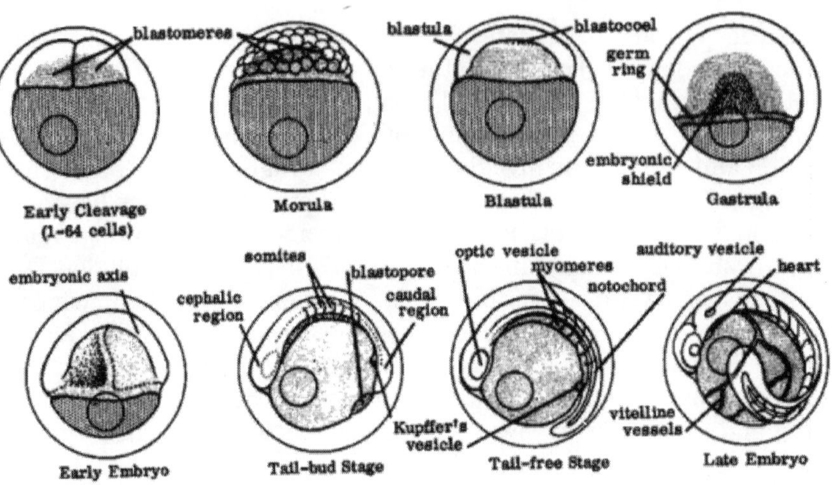

Fig 4.5 - Embryonic Development in Fish

1.10. Follicular atresia

All the oocytes developing in the ovary of a fish do not reach full maturity and several of them degenerate. These are called atretic oocytes or pre-ovulation corpora lutea and are seen in a section of the ovary during the pre-spawning and after-spawning periods. Several oocytes are thus lost from the ovary due to degeneration of the oocytes at various stages of maturation. This influences the fecundity or the reproductive potential of the fish. The process of atresia has been seen in some teleosts such as *Mystus seenghala, Wallago attu, Channa punctatus, Tor sp, Clarius batrachus, H. fossilis* etc. Both mature and young oocytes become atretic and the follicular cells surrounding them cannot produce healthy mature ova.

1.10.1. Causes of follicular atresia

Various external and internal factors are believed to be responsible for the degeneration of oocytes. These are hormones, temperature, photoperiod, overcrowding, nutritional state of the breeders, physico-chemical nature of water, etc. Studies show that an imbalance of hormones and steroids causes degeneration of oocytes. Various environmental factors, such as light, temperature, pH, salinity etc, must be favourable for healthy growth of oocytes. Hence, any disturbance in these factors may cause atresia.

In females, gamete development also involves the manufacture and incorporation of the yolk sac into the developing oocytes (vitellogenesis). During the early part of gonadal development, the oogonia becomes associated with a number of pre-follicular cells, and it is from these that the ovarian follicle is derived. In the salmonids, the oocytes increase in diameter from about 50 μm to 500-1000 μm before the start of vitellogenesis. The incorporation of yolk that occurs during the vitellogenic phase leads to the diameter of the oocyte increasing to around 5000 μm (5 mm). Thus, there is an enormous increase in the oocyte volume during vitellogenesis. Most of this increase is due to the incorporation of exogenously synthesized yolk into the oocyte's cytoplasm.

The yolk is derived from vitellogenin, a lipophospho-protein- Ca complex. Vitellogenin is synthesized in the liver and released into the blood system. The vitellogenin is transported to the ovary in the blood and set aside by the oocyte. Following the uptake into the oocyte, the vitellogenin is spilt into 2 major components- phosvitin and lipovitellin, which form the yolk stores.

1.11. Modes of reproduction

1.11.1. Live bearers/ internal bearers

Most sharks give birth to live young, but some release eggs that hatch later. The gestation period (the time the embryo spends inside the female, developing) can be almost two years long (the spiny dogfish shark has the most prolonged shark gestation period).

Baby sharks (called pups) are born with a complete set of teeth and are fully ready to care for themselves. They quickly swim away, even from their mothers who might eat them. Litter size ranges from one or two pups (for a great white shark) to over 100 (for a large blue and whale shark).

i. Ovi-ovoviviparous

This is considered to be the first evolutionary step towards live-bearing. In some sharks and rays, after internal fertilization and after several weeks of incubation, numerous embryos are laid, each covered with a horny capsule. Among teleosts, they lay their embryos immediately after fertilization.

ii. Ovo-viviparous or Aplacental Viviparity

This is best known from squaliform sharks and the coelacanth (Latimeria). In these fishes, development is internal. The eggs hatch, and the babies develop inside the female's body, but no placenta nourishes the pups. The large active young are born after the yolk sac has been absorbed or have their young developed inside them without direct maternal nourishment.

The eggs develop in the oviduct of elasmobranchs and the ovary of teleosts. The gestation period can last from a day or 2 in small tropical teleosts to 1 or 2 years in sharks. The pups eat any unfertilized eggs and each other (they are oviphagous). Very few pups in a litter survive until birth due to this form of sibling cannibalism. E.g., for teleosts are Scorpaneidae, Cottidae, and Hexagrammidae.

iii. Viviparous

To be precise, the female keeps her developing eggs in the final section of the oviduct, called the womb or uterus and the babies are fed by a placenta, which transfers nourishment from the mother to the babies (via an umbilical cord, which is connected to the baby shark behind the pectoral fins). The placenta helps transfer nutrients and oxygen from the mother's bloodstream and transfers waste products from the baby to the mother for elimination. Following a long gestation period, gives birth to fully developed, highly independent pups. In other words, sharks give birth to living offspring. The long gestation period inside the mother's body protects the young from enemies. The number of pups in a litter range from 2-20 or more.

E.g., Hammerhead sharks (Sphyrinidae), Carcharhnidae, Embiotocidae, Poecillidae.

1.11.2. Oviparous, or "egg-laying"

Lay undeveloped eggs, external fertilization (90% of bony fish), internal fertilization (some sharks and rays).

In fishes, oviparity is the most common. The adult can produce many offspring, which they broadcast into the plankton column. When the offspring settle out of the plankton, they may be in new environments, allowing for a great area where the young may survive. This mode also has disadvantages; when born, the fish must first go through a larval stage for growth before they transform into the adult stage. They may not find a suitable environment when they settle out of the plankton column. The survival of individual eggs is meagre, so millions of eggs must be produced for the parent to produce offspring successfully.

The other modes have their advantages; the eggs are much less prone to predation when carried within the mother, and the young are born fully advanced and ready to deal with the environment as miniature adults. The adult must supply nutrients to its offspring and can only produce a few eggs at a time. The youngones are limited to their parents' environment, and if this environment is deteriorating, they are stuck with it.

The sharks deposit eggs in the ocean, which will hatch later if predators do not eat them. Either parent does not guard the eggs. Shark eggs (sometimes called "mermaid's purses") are covered by a tough, leathery membrane. Their shape ranges from pouch-like to screw-shaped. Some eggs (like those of cat sharks) have tendrils that attach the egg to objects on the sea bed. The egg has a yolk that feeds the embryo, much like a chicken egg. Oviparous sharks include the Zebra shark, the catshark, the swell shark, the necklace carpet shark, some Epaulette sharks, and the Horn shark.

1.12. Seasonal regulation of reproduction/ environmental influence on the reproductive cycle

The general cues that trigger gamete growth and development appear to be environmental. These environmental cues are often linked to the annual cycles of daylength and temperature variations. For e.g., *Cyprinus carpio* and many other cyprinids, the gonadal maturation process is thought to be triggered by a combination of increasing daylength and the commencement of the rise in water temperature. The spawning season comes to an end during the summer. It is thought that the cessation of spawning activity may partly be due to the inhibitory effects of high summer water temperatures on gonadal differentiation and maturation. Spawning resumes as temperature begins to fall with the approach of winter. This indicates that temperature is not the only environmental factor exerting a controlling influence over reproduction. Thus, it is thought that photoperiod and temperature are major environmental cues responsible for mediating the reproductive cycle in several species of fish that inhibit temperate latitudes.

In the tropics, only very minor seasonal changes in photoperiod and water temperature may exist. Consequently, these 2 environmental factors are unlikely to act as major cues influencing the reproductive cycle and the timing and duration of the spawning season. In some tropical species, there does not appear to be any distinct spawning season, and members of the population can be found to be engaged in reproductive activities at almost any time of the year. In other species, spawning activities do seem to vary on a seasonal basis, with factors such as variations in rainfall possibly acting as environmental cues that trigger changes in gonadal development and maturation.

1.13. Hormonal control of reproduction

Hormones of the pituitary gland play a significant role in reproduction. The pituitary releases gonadotropin hormones, which control the growth of testes and ovaries in fishes. The pituitary also shows many indirect effects on behaviour by their participation in developing secondary sexual characteristics in fishes.

However, the hormones of the gonads and adenohypophysis of the pituitary (anterior pituitary) play a significant role in controlling reproductive processes. Some evidence proves that neurohypophysial and thyroid hormones are also involved in regulating specific components of reproductive behaviour in fish. Other hormones indirectly affect reproductive behaviour by their effects on metabolism and growth.

The primary hormones that regulate sex and reproduction are sex steroids such as androgens, estrogens and progesterone, which have masculinizing, feminizing and gestational effects, respectively. These hormones are secreted from the target glands, gonads, which are under pituitary control. The pituitary gland manufactures many hormones, and their mechanism of action varies considerably.

The gonadotropins are secreted from the anterior pituitary (adenohypophysis) and affect their target glands- the gonads.

1.14. Shellfish

The reproduction of crustaceans also varies. Some change sexes in the course of their lives; others are even hatched with both sexual organs.

1.14.1. Separate Sexes

The sexes are separate in most crustacea and usually can be differentiated by secondary sex characters. Chief among these characters are the size and shape of the body, appendages, or both, and placement of the genital apertures. For these crustaceans, reproduction functions similarly to many of the fish species. The female produces eggs in the ovaries that are either released in the water or attached to the female, waiting for fertilization. The male then fertilizes the eggs. Lobsters perform serial monogamy, meaning the males are engaged with only one female at a time but will repeat the fertilization with many females. The female sheds her outer shell and moves into the male's lair for approximately two weeks.

1.14.2. Hermaphroditism

A hermaphrodite is an animal or plant with both male and female reproductive organs. In crustaceans such as the Notostraca or tadpole shrimp, shrimp are hatched, containing both male and female sexual characteristics. The ovaries contain sperm-producing lobes and eggs. In other species, such as the cirripede, males and hermaphrodites exist simultaneously. The males, if present, are much smaller than the hermaphrodites and called dwarfs. They attach themselves to the hermaphrodite to assist in the fertilization of the eggs.

1.14.3. Sex Change

In some shrimp species, it is common for them to change sexes as they age. The Aesop Shrimp (*Pandalus montagui*) start as males and change to females after 13 months. They are found in the Atlantic Ocean. The Oregon Pill bug (*Gnorimosphaeroma oregonense*) is the only crustaceans that hatch as females and, after maturity, become males.

1.15. Reproduction in penaeids

Penaeids are dioecious, and the external structures of the genital system are the major dimorphic features. The male has two pairs of modified abdominal appendages (Pleopods) on the first and second abdominal segments (the **petasma** and **appendix masculina**) that deliver sperm to the female's external receptacle (the **thelycum**) located between the bases of the fifth walking legs (pereopods). The petasma, appendix masculine and thelycum are located on the ventral surface.

The petasma is formed by the endopodites of the first pair of pleopods, which are modified into interlocking structures for spermatophore transfer. The appendix masculina are on the endopodites of the second pair of pleopods and serve to separate the petasma into two component halves. The thelycum may be "open" or "closed", depending on the species. "Closed" thelyca are those where a male places the spermatophore in the groove below the plates while the female is in the soft exoskeleton stage following molting. The spermatophore is stored for some time before spawning. "Open" thelyca are not enclosed by plates, and the spermatophore must be placed on it by a male when the female's exoskeleton is hard, usually within hours of spawning. The presence of a spermatophore on the female is evidence that she has successfully mated. Open thelycum are found in some shrimp species endemic to the Western Hemisphere, such as *P. stylirostris* and *P. vannamei;* while closed thelycum are characteristic of most Asian species, such as *P. monodon, P. chinensis, P. indicus* and *P. merguiensis.*

Internal organs of the male reproductive system include paired testes, vas deferens and terminal ampoules for spermatophore storage. The female reproductive system includes paired (but partially fused) ovaries that extend from the mid-thorax to the posterior end of the abdomen and oviducts terminating adjacent to a single thelycum.

Dioecious species are those whose individual members can usually produce only one type of gamete; each organism belonging to a dioecious species is distinctly female or male (the word deriving from the Greek for "two households"). Most animal species (for example, all mammals and most reptiles) are dioecious.

Crustaceans produce eggs, which will be fertilized by sperm in the same manner as other animals. The eggs are produced in the ovaries of the female and passed to the outside through oviducts. The sperms are produced in tubular testes in the male. After eggs have been fertilized, they begin development and then hatch.

The eggs of most crustaceans are carried attached to the female until they hatch. Some females develop brood pouches in which the young are retained. A nutrient secretion that sustains the young until they are released. Penaeid shrimp and a few of the lower crustacea deposit their eggs in the medium they live and, in some cases, deposit them in aquatic vegetation.

Hatching eggs can take several days to weeks, depending on the species, for the young larvae to get detached. From this point, they are on their own and must feed, grow, swim and survive. After a series of transformations, the larva becomes a miniature adult.

1.15.1. Life histories vary from the simple to the complex

Mating takes place in shallow offshore waters and involves the transfer of spermatophores from the male to the female shrimp. Actual spawning takes place in deeper offshore waters. The eggs are released and fertilized externally in the water. During spawning, the last 3 pairs of periopods are held tightly together and are flapped vigorously. The unfertilized eggs are extruded from the paired ovipores at the base of the third periopods. Simultaneously, stored sperm are released from the thelycum. Both are ejected with considerable force, where the eggs appear greenish and the sperm whitish, with clouds exiting beneath the females as they swim. The sperm are non-motile and must come into contact by passive collision with the egg's surface at the moment of ejection.

The vigorous flapping of the female's pereiopods is believed to create an intense vortex that increases the likelihood of sperm and egg coming into contact. A female may lay between one-half to one million eggs at a single spawning. Within 24 hours, the tiny eggs hatch into a microscopic larva known as a nauplius. The protozoeal, mysis, and postlarval stages follow the nauplius stage. Passage from the nauplius to the postlarval stage takes several weeks. All of the developmental stages are found in the offshore plankton. The larvae are capable of little horizontal, directional movement and cannot swim independently of the water currents. They are reliant upon favourable currents to transport them to inshore waters. The migration from offshore waters to coastal bays occurs during the last planktonic stage, and shrimp enter estuarine nursery grounds as postlarvae. Once they move into brackish waters, the postlarvae abandon their planktonic way of life and become part of the benthic community.

Postlarval and juvenile shrimp occupy the shallow, brackish waters of the Sound, where they feed and grow. Growth of the young is rapid when the waters are warm (above 20^{0}C). Young shrimp remain in the estuary until they approach maturity. Adult shrimp migrate offshore to spawn, and the cycle is repeated.

1.16. Decapod Crabs - Reproduction and Development

Decapod crabs are built like folded-up lobsters. The abdomen, equivalent to a tail, is folded up tightly underneath its body to form an abdominal flap.

When you turn a crab over and look underneath, you can tell a crab's sex by looking at its abdominal flap. A male crab has a **small triangular flap**, while a female crab has a broad, **oval-shaped abdominal flap**.

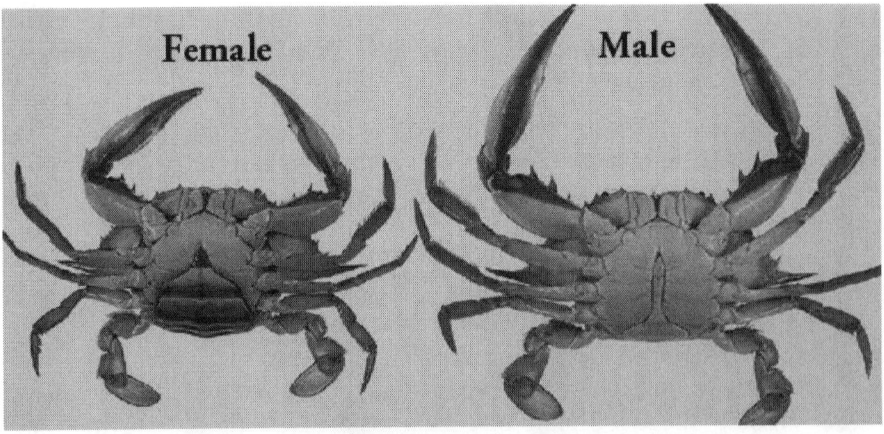

Fig 4.6 - Abdominal flap of female and male crab

When a male and female crab mate, many female decapod crabs can store the male sperm until their eggs are ready to be released. When the eggs are released, the stored sperm flows over them, fertilising them.

The reproduction habits of crabs are attractive as well. The male delivers his sperm to the female for egg fertilization via the gonophores located on the legs. Then, they find a comfortable place for the female to lay her brood and watch over them until they hatch. The first two sets of appendages are used for sperm transfer; therefore, the gonophores are located more towards the body's midline.

The female crab holds the fertilised eggs in a big spongy mass between its abdominal flap and the body. The eggs are cemented to the pleopods, which are tiny legs, creating the "berried" appearance.

The female crab continually "waves" water over the eggs with the pleopods to keep the eggs healthy. When the eggs hatch into **zoea larvae**, they drift away in the ocean currents as plankton. As the juvenile crab grows, it goes through a series of moults, each larval stage changing form and function as it grows in a metamorphosis process.

More segments are added to the end (posterior) at each moult, and the clawed limbs replace the feathered limbs. The **megalops** stage more closely resembles the adult decapod crab.

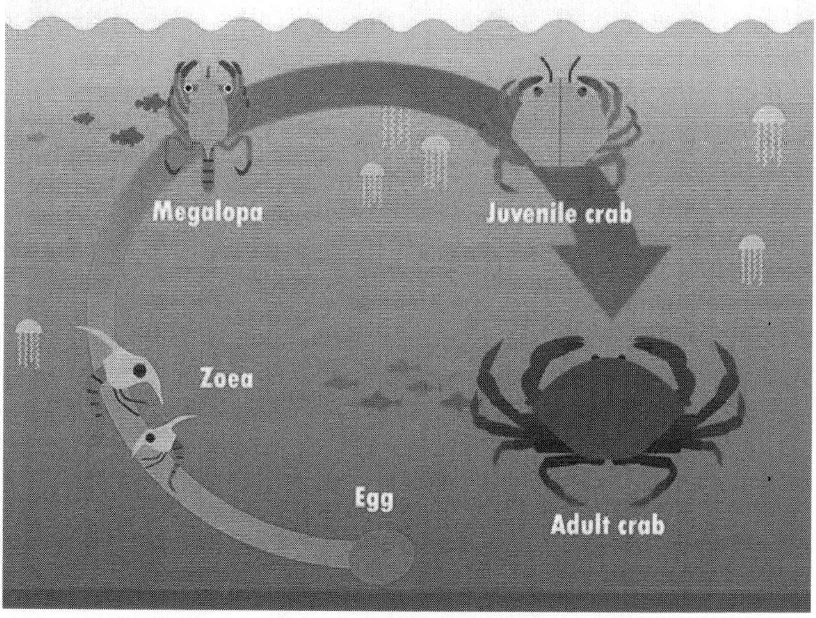

Fig 4.7 – Life cycle of crab

1.17. Reproduction in Molluscs

Molluscs are primarily of separate sexes, and the reproductive organs (gonads) are simple. Reproduction via an unfertilized gamete (parthenogenesis) is also found among gastropods of the subclass Prosobranchia. Most reproduction, however, is by sexual means. Eggs and sperm are released into the water and fertilization occurs there. In prosobranch gastropods, water currents may cause a simple internal fertilization within the mantle cavity, or males may fertilize eggs internally using a muscular penis. Both male and female reproductive organs may be present in one individual (hermaphroditism) in some species, and various groups exhibit different adaptations to this body form. For example, in hermaphroditic bivalves and prosobranch gastropods, male and female gonads are functional at separate times and in rhythmic and consecutive patterns (successive hermaphroditism). Conversely, male and female gonads are functional at the same time (simultaneous hermaphroditism) in solenogasters and many other gastropods.

Fertilization occurs by the transfer of capsules containing sperm (spermatophores) in cephalopods and some gastropods. In cephalopods, transfer of spermatophores is usually combined with copulation by a modified arm or hectocotylus. Copulation in solenogasters, often utilizing a special genital cone, may be supported by copulatory stylets. Various penis formations, in part with copulatory stylets, or darts, are widely found in gastropods.

Eggs are deposited singly or in groups, generally on some hard surface and often within jelly masses or leathery capsules. Squids of the suborder Oegopsida and some gastropods have eggs that are suspended in the water. Fertilized eggs commonly undergo spiral cleavage, as in annelids and several other "protostome" phyla. On the other hand, the eggs of cephalopods possess a large amount of yolk, displacing the dividing cells and causing a characteristic type of development.

Many molluscs develop into free-swimming larvae, either feeding (planktotrophic) or nonfeeding (lecithotrophic). The larva in primitive bivalves is a pericalymma (test cell) larva in which the embryo is protected below a covering (test) of cells provided with one to four girdles of cilia, at the apex of which is a sensory plate of ciliated cells. After the developing juvenile has grown out apically of the test (which then is lost), the animal settles and develops into an adult. The test in other lecithotrophic larvae is restricted to a preoral girdle of ciliated cells (the prototrochus) called the trochophore larva. Trochophores are encountered in developing many marine annelid species (phylum Annelida). The trochophore larva develops into a veliger larva in more advanced molluscs (such as in marine gastropods and bivalves). In these generally planktotrophic larvae, the girdle of ciliated cells widens to form a velum that entraps food and propels the microscopic mollusc through the water. As the larva develops, the shell, mantle cavity, tentacles, and foot appear. After a specific time, which varies according to species and environmental conditions, the larva loses the velum and metamorphoses into an adult. A substantial change in shell morphology usually marks the transition to adult form.

Secondary (newly evolved) larvae have developed among some freshwater bivalves and some cephalopods. Maternal protection of the developing eggs (brood) is not unexceptional behaviour in solenogasters, bivalves, and certain gastropod adults. Direct development without a larval stage or the bearing of live young from a yolky egg, or both, are typical in cephalopods and most nonmarine (and many marine) gastropods. Many species go through two breeding seasons per year, whereas in some cephalopod species, mating or egg-laying appears to be rapidly followed by death effected by hormones.

1.18. Reproduction in Bivalvia (Pelecypoda)

The majority of bivalves are dioecious (two sexes). Their two gonads are closely situated next to each other, encompassing the intestinal loops. The gonoducts are very simple, as there is no copulation among bivalves.

In the Protobranchs and Filibranchs, the gonoducts open directly into the nephridia and provide for the exit of sperm and eggs.

In the Eumellibranchs, the gonoducts open directly into the mantle cavity very close to the nephridiopore.

A few bivalves, such as the Cockles (*Cardiidae*), *Poromyidae*, a few of the oysters and scallops (*Pectinidae*), and some of the fresh water clams *Sphaeriidae* and *Unionidae* are hermaphroditic (one sex).

In most bivalves, sperm and eggs are released into the surrounding water, where fertilization occurs. The eggs and sperm deposited into the suprabrachial chamber are swept out along with the exhalent current.

In a few bivalves, such as the common oyster (Ostrea edulis), fertilization occurs within the suprabranchial chamber when sperm is drawn in along with the inhalent current. The fertilized eggs then develop in the gill filaments.

In some of the freshwater hermaphrodites, self-fertilization may actually occur in the genital ducts before the eggs are deposited into the suprabranchial chamber. The eggs then travel into the water tubes of the gill, and they develop into larvae.

1.19. Reproduction in Cephalopoda

Cephalopods are usually dioecious (two sexes), and fertilization is internal. Courtship, which can be quite elaborate in some species, is often a precursor to copulation. Gonads are located in the posterior of the body.

In some species, males can be distinguished by the modified sucker discs found at the tips of their longer two tentacles. The male uses these long arms to remove a sperm packet from his mantle cavity and then insert this packet into the female's mantle cavity. This arm is autonomous and sometimes can get broken off while mating, but it is retained within the female's cavity. If this happens, the male grows a new one: no problem!

Within two months of mating, the female octopus will attach long strands of clustered eggs (resembling a cluster of grapes) to the ceiling. While the eggs are incubating, the female gently caresses them to keep them clean and bacteria-free. She also keeps fresh, oxygenated water flowing over her precious eggs. When they are ready to hatch, her caresses become more rigorous, which helps the young escape from their egg sacs.

The pelagic Cephalopod family Argonautidae, commonly known as the paper nautilus, remarkably adapted to egg deposition. The two dorsal arms of the female are greatly expanded at their tip to form a membrane. The expanded portion of each arm secretes one-half of a beautiful calcareous bivalved shell. She then deposits her eggs directly into this case. The shell acts as both a brood chamber and a retreat for the female. The posterior of the female usually remains in this shell. The male, much smaller than the female, does not have such a shell and is often found in cohabitation within the same shell, more or less as a freeloader.

The gonads, ovaries and testes of all vertebrates are similar in structure and function. The gonads are usually paired (although they may be fused or single in some groups). The testes in all vertebrates produce millions of sperm at a time. The major differences in male reproductive systems involve the mechanisms for transferring those millions of sperm to the ova.

Ovaries produce ova surrounded by fluid filled sacs called follicles. These follicles burst releasing ova into the coelom and then the oviduct. The number of ova produced at any one time varies considerably depending upon the habitat of the animal and the amount of parental care the youngones will receive. Thus, fish like the cod, which release unprotected eggs into the ocean and do not care for their young, may simultaneously produce and release more than a million eggs. Other fish and amphibians that do not broadcast their ova as far afield or give some measure of parental protection to the eggs may produce thousands at once. Birds and mammals, which invest considerable parental energy in protecting their offspring before and after birth, produce only a few eggs (ova) at once.

Once ova are released from the ovary, they are transported through the oviduct toward the exterior. In the case of fish and amphibians, which release shell-less eggs, the oviduct is essentially an unspecialized tube leading indirectly to the exterior. Birds and reptiles produce shelled eggs; the shell is critical in protecting the egg from desiccation in the terrestrial environment. The oviducts in these animals are modified, closely associated with a shell-forming gland and capable of considerable distension to accommodate the enormous, inflexible eggs that result.

CHAPTER 2

STRUCTURE AND FUNCTION OF IMPORTANT ENDOCRINE GLANDS

2.1. Introduction

A **gland** is a group of cells that produces and secretes or gives off chemicals. A gland selects and removes materials from the blood, processes them and secretes the finished chemical product for use somewhere in the body. Some types of glands release their secretions in specific areas. There are 2 types of glands.

1. Exocrine glands
2. Endocrine glands

2.2. Exocrine glands

Exocrine glands are glands that secrete their products (enzymes excluding hormones and other chemical messengers like neurotransmitter - communicates to adjacent cells, neuropeptide - a protein sequence which acts as a hormone or neurotransmitter, pheromone - a chemical factor that triggers a social response in members of the same species) into ducts (duct glands- Any tube or canal by which a fluid or other substance is conducted or conveyed or one of the vessels of an animal body by which the products of glandular secretion are conveyed to their destination). They are the counterparts to endocrine glands, which secrete their products (hormones) directly into the bloodstream (ductless glands). **Exocrine glands** release their secretions through ducts into the external environment or directly to the exterior surface or an area contiguous with the exterior surface, not utilizing the blood to deliver the product. Most utilize ducts, but some glands are classified as exocrine glands, such as the digestive glands in the intestine and the sebaceous glands of the skin (The **sebaceous glands** are microscopic glands in the skin that secrete an oily/waxy matter called **sebum**, to lubricate the skin and hair of mammals) open directly onto the body surface. Typical exocrine glands include sweat glands, salivary glands, stomach, liver and pancreas. The chemical substances released by exocrine glands include sweat, digestive enzymes and tears (through tear ducts).

2.3. Endocrine glands

Endocrine glands, on the other hand, release or secrete more than 20 major hormones or chemicals directly into the bloodstream, where they can be transported to cells in other parts of the body. The endocrine system consists of a set of glands which are ductless. Hormones are chemical messengers. Hormones play an essential role in the chemical coordination of the animal body. In fishes, many endocrine glands are present, which secrete many hormones and control various body processes. The endocrine system coordinates the body's activities with the nervous system. The endocrine system is slower than the nervous system because hormones must travel through the circulatory system to reach their target. However, the nervous system conducts signals much quicker than the endocrine system.

Chemicals that interfere with the normal functioning of this complex system are known as "endocrine disruptors." Disruption of the endocrine system can occur in various ways.

For example, some chemicals may mimic a natural hormone, fooling the body into over-responding to the hormone. Other chemicals may block the effects of a hormone in parts of the body normally sensitive to it. Still others may directly stimulate or inhibit the endocrine system, leading to overproduction or underproduction of hormones.

The major endocrine glands

1. Hypothalamus
2. The pituitary gland
3. The thyroid gland
4. Adrenal cortical tissue or Inter-renal tissue
5. Chromaffin tissue
6. The ultimo-branchial gland
7. Islets of langerhans or Endocrine Pancreas
8. The corpuscles of stannius
9. The urophysis (Caudal neurosecretory System)
10. Gonads
11. Pineal gland

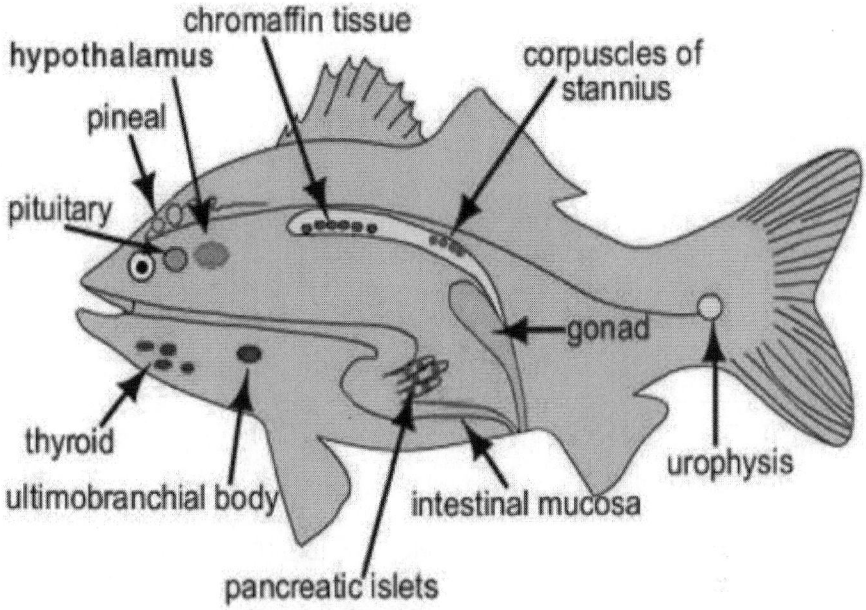

Fig 4.8 - Major endocrine glands

2.3.1. Hypothalamus

The **Hypothalamus** is a portion of the diencephalon that lies inferior (below) to the thalamus and contains several small nuclei with various functions. One of the most critical functions of the hypothalamus is to link the nervous system to the endocrine system via the pituitary gland (hypophysis). The hypothalamus drives the endocrine system.

The brain is made of three main parts: the fore brain, mid brain, and hind brain. The forebrain consists of the cerebrum, thalamus, and hypothalamus (part of the limbic system). The mid brain consists of the tectum and tegmentum. The hind brain is made of the cerebellum, pons and medulla. The midbrain, pons and medulla are often called the brainstem.

The thalamus is the largest structure in the diencephalon, the part of the brain situated between the mid-brain (mesencephalon) and fore brain (telencephalon).

Function of thalamus:

- Motor Control
- Receives Auditory, Somatosensory and Visual Sensory Signals
- Relays Sensory Signals to the Cerebral Cortex

It synthesizes and secretes certain neurohormones, often called hypothalamic-releasing hormones, that travel in blood vessels to the anterior pituitary and these, in turn, stimulate or inhibit the secretion of pituitary hormones. The blood vessel that carries hypothalamic-releasing hormones from the hypothalamus to the pituitary is called a ***portal vein*** because it connects two capillary beds. One capillary bed is in the hypothalamus, and the other is in the anterior pituitary. Release-inhibiting hormones produced by the hypothalamus inhibit the pituitary from secreting its hormones.

The hypothalamus is responsible for specific metabolic processes and other activities like maintaining homeostasis (constant internal conditions) by regulating the internal environment (examples: heart rate, body temperature, water balance, hunger, thirst, fatigue, Sleep, circadian cycles and the secretions of the pituitary gland).

2.3.2 Pituitary gland

The pituitary, or the hypophysis, is the most important endocrine gland in teleosts and consists of 2 main parts or components: the neurohypophysis (the posterior pituitary), neural part and the adenohypophysis (the anterior pituitary), a glandular part. Neurohypophysis is derived from the brain, i.e., from the floor of the embryonic diencephalon (forebrain), and the adenohypophysis develops as an ectodermal evagination of the roof of the buccal cavity. The pituitary gland is present in a shallow depression. The sella tunica, which is found in the sphenoidal area of the skull, is situated ventral to the brain in the region of diencephalon just behind the optic chiasma (formed by the decussation of the optic nerves) and is connected or functionally linked to the hypothalamus by the pituitary stalk (also named the "infundibular stem", or simply the "infundibulum") which may be absent in some species. It is from the hypothalamus that hypothalamic tropic factors are released to descend the pituitary stalk to the pituitary gland, where they stimulate the release of pituitary hormones. The pituitary gland is the 'master' endocrine gland because it controls other glands. Nevertheless, both of the lobes are under the control of the hypothalamus; the anterior pituitary receives its signals from the parvocellular neurons (**Parvocellular cells**, also called **P-cells**, are neurons located within the parvocellular layers of the lateral geniculate nucleus (LGN) of the thalamus) and the posterior pituitary receives its signals from magnocellular neurons (**Magnocellular neurosecretory cells** are large cells within the supraoptic nucleus and paraventricular nucleus of the hypothalamus).

Anterior pituitary (Adenohypophysis)

This is made of 3 parts: the rostral pars distalis, proximal pars distalis and the pars intermedia.

Rostral pars distalis is the anterior lobe secrets Prolactin, Adrenocorticotrophic hormone, Thyroid stimulating hormone.

Proximal pars distalis is the posterior lobe secrets follicle-stimulating hormone (GTH I), Luteinizing hormone (GTH II), Growth hormone.

Pars intermedia secretes melanocyte-stimulating hormone.

Somatotrophs (STH cells) or growth hormone (GH): These hormones are secreted by somatotroph cells and are mainly found in pars distalis and are released under the influence of hypothalamic Growth Hormone Releasing Hormone (GHRH) or inhibited by hypothalamic Somatostatin. These hormones increase the fish's growth rate and basal metabolic rate. Growth is retarded in a hypophysectomised (Hypophysectomy or hypophysis is the removal of the pituitary gland) fish, and injection of GH in such a fish causes an increase in length. It is believed that an increase in the length of teleosts by GH is caused by increasing appetite and protein synthesis and by stimulating the synthesis and release of insulin.

Prolactin (PRL): Prolactin (PRL) is also known as the 'Luteotropic' hormone (LTH). It is released by lactotroph cells under the influence of multiple hypothalamic Prolactin Releasing Factors (PRH). These cells secrete prolactin hormone, which is involved in osmoregulation and influences the movement of water and sodium through the gills, kidneys, and urinary bladder. It is also concerned with regulating the level of sodium in blood plasma. This hormone is also believed to influence lipid metabolism and storage of fat. The thyroxine level of serum is reduced by prolactin. A prolactin-releasing inhibiting hormone (PIH) is reported to control the secretion of prolactin in fish.

Melanocyte-Stimulating Hormone (MSH): This influences the pigments of the skin, i.e., the darkening of the skin in fish. **(Hormones of Pars intermedia-** 2 hormones Melanophore Dispersing Hormone (MDH) and Melanophore Concentrating Hormone (MCH) are secreted in this region. These influence the pigments of the skin, i.e., darkening of the skin in fishes).

Thyrotrophs or Thyroid Stimulating Hormones (TSH): It is released by the thyrotroph cell of adenohypophysis under the influence of hypothalamic thyrotropin-releasing hormone (TRH). It stimulates the thyroid gland to secrete thyroxine. TSH is, in turn, controlled by thyroxine. It also controls the overall metabolic process of the body.

Gonadotrophs (FSH and LH cells): There are two 'Gonadotropins' that are released by the gonadotroph cell. The gonadotrophic hormones of the pituitary gland appear to be regulated by the gonadotrophic releasing hormone (GnRH) identified in different locations of the teleostean brain. Besides, a gonadotrophic inhibiting hormone (GIH) is also present in the brain of teleosts. **LH** and **FSH** from the anterior pituitary stimulate the gonads (ovaries and testes). Follicle cell-stimulating hormone controls the development of follicles in fishes and sperm formation. Leutinizing hormone (also referred to as 'Lutropin' or 'LH' or in males, 'Interstitial Cell Stimulating Hormone' (ICSH) controls ovulation and spermatogenesis. Both are released under the influence of Gonadotropin Releasing Hormone (GnRH).

Corticotrophs or Adrenal Corticotrophic Hormones (ACTH) are released by corticotroph cells under the hypothalamic Corticotropin Releasing Hormone (CRH). It controls the secretion of the adrenal gland. Some investigators have found that the adrenocorticotropic hormone (ACTH) is also under the control of the hypothalamus, and a corticotrophin-releasing factor (CRF) is present in the extracts of the telencephalon (The anterior portion of the forebrain).

Posterior pituitary (Neurohypophysis)

The neurohypophysis is also called Pars nervosa. It secretes Vassopressin and oxytocin

1. Vasopressin hormone

It regulates osmoregulation in fishes

2. Oxytocin

It controls mating in fishes and the process of egg-laying.

Most of the antidiuretic hormone ADH, also known as vasopressin and AVP arginine vasopressin, is released from the supraoptic nucleus in the hypothalamus and stored in the neurohypophysis. It regulates Oxytocin, most of which is released from the paraventricular nucleus in the hypothalamus.

Neuro-endocrine regulation of the pituitary gland

The pituitary gland controls several metabolic activities of the fish through its hormones, but the brain's hypothalamus regulates the secretion of its hormones. Studies have shown that the nucleus preopticus (NPO) and the nucleus lateralis tuberis (NLT) are the critical nuclear areas that control the pituitary's activities. The NPO is situated dorsal to the optic chiasma, and its axons (An **axon** is a long, slender projection of a nerve cell or neuron that conducts electrical impulses away from the neuron's cell body) extend into the pituitary gland. The NLT lies along the ventral surface of the infundibulum, and its neurons innervate the hypophysis cells. The hypothalamus controls the activities of the pituitary gland either through these axons directly or by transferring neuro-endocrine hormones through a hypothalami-hypophysial blood portal system.

Extracts of the pituitary gland are extensively used in fish-induced breeding in fish culture.

2.3.3. The thyroid gland

The thyroid gland in teleosts is present in a diffused condition around the ventral aorta and the afferent branchial arteries of the gills. In general, it is located below the gills and is not always divided into distinct lobes. However, in some teleosts, patches of thyroid tissue are found elsewhere in the body and are associated with the kidneys, spleen, heart, or eyes.

Under a microscope, a large number of thyroid follicles **(discrete, cyst-like units of the thyroid gland lined with cuboidal epithelium and** filled with a colloid substance) is seen in the thyroid tissue. The thyroid follicles are very similar to mammalian thyroid tissue. Thyroid follicles vary in shape and size, are bound together by connective tissue and are richly supplied with blood capillaries. Thyroid follicles are distributed throughout the connective tissue of the pharyngeal area and may be observed around the eye, ventral aorta, hepatic veins and anterior kidney. It is essential to realize that thyroid tissue can be widely distributed. Temperature and photoperiod have been reported to exert a specific influence on the structure and function of the thyroid in teleosts.

Functions

Thyroid follicles synthesize thyroid hormone after extracting iodine from the blood and combining it with tyrosine (an amino acid). The hormone in fish contains mono-iodotyrosine (MIT), diiodotyrosine (DIT), triiodothyronine (T_3) and thyroxine (T_4). The secretion of hormones from the thyroid is controlled by the pituitary gland's thyrotrophic stimulating hormone (TSH), which performs essential functions in fish metabolism. The synthesis of thyroid hormones is tightly controlled. When the amount of thyroid hormone in your blood drops, the pituitary gland secretes a hormone called thyroid-stimulating hormone (TSH). As its name suggests, TSH then stimulates the thyroid gland to increase its uptake of iodine from the blood to synthesize more thyroxine (T4). When necessary, thyroxine is converted to the metabolically active triiodothyronine (T3), which involves removing one iodine atom from T_4.

Surgical thyroidectomy is impossible in teleosts as the thyroid is diffused and follicles are scattered in many organs. However, the functions of this gland have been extensively studied utilizing chemical thyroidectomy (using thiourea) and radio thyroidectomy using I^{131}. Despite the efforts of several investigators, the functions of thyroid hormone have not been fully understood.

Thiourea is an organosulfur compound. It is structurally similar to urea, except that a sulfur atom replaces the oxygen atom, but the properties of urea and thiourea differ significantly.

Experiments suggest that the thyroid gland is involved in several physiological activities of the fish, such as growth, osmoregulation, carbohydrate metabolism, reproduction and migration.

It plays a vital role in oxygen consumption. Gold fish treated with thiourea exhibit decreased oxygen consumption. However, the definite influence of thyroid hormone on oxygen consumption could not be established by surgical thyroidectomy and treatment of the fish with various chemicals and hormones.

The thyroid influences the growth of fish. Fish treated with thyroid hormone by immersion or injection show an increase in length and weight, possibly due to enhanced growth hormone production. Surgical thyroid-ectomy or radio thyroid-ectomy in salmon and trout retarded their growth.

Thyroid appears to be involved in the carbohydrate metabolism of fish. When the gland is active, liver glycogen is reduced. Treatment of fish with thiourea also reduces glycogen in the liver.

Thyroid secretion appears to play an important role in reproduction and migration of fish. Treatment with thyroxine stimulates ovarian development, probably through the pituitary gland.

In brown trout and rainbow trout, T_3 (***triiodothyronine*** because it contains 3 iodine atoms) is essential for adaptation to seawater.

Thyroxine also promotes maturation in fishes. Scale and bone formation in fish is also controlled by thyroxine. Both T_4 and T_3 have similar effects on target cells. In most target tissues, T_4 is converted to T_3.

2.3.4. The Adrenal Glands

Adrenal cortical tissue or Inter-renal tissue

No true adrenal gland is present in most fish (except sculpins- The sculpin is a spiny fish in the order Scorpaenidae, family Cottidae). They are relatives of scorpion fish and are common in North American waters. The sculpin can be found in salt and fresh water, commonly in tide pools and other shallow water, where it clings close to the bottom). In many teleosts, the inter-renal and chromaffin cells are found in the head kidney, composed of haemopoietic or lymphoidal tissue.

Only a few renal tubules are seen in the head kidney. The inter-renal cells may be located in one or several layers along the post-cardinal vein. In *Puntius ticto*, these cells form a thick glandular mass, but in *Channa punctatus*, they are arranged in cords of cells along the vein. Considerable variations have been seen in the structure and distribution of inter-renals in teleosts, and fishes can be divided into 4 groups.

Group I: Inter-renal cells surround the post-cardinal vein or its large branches

Group II: Inter-renal cells are diffused throughout the head kidney along small and medium-sized veins.

Group III: Inter-renal cells are associated with venous sinuses in the head kidney

Group IV: Inter-renal tissue forms a solid mass in a localised area in the head kidney

The inter-renal cells may be columnar, polygonal, cuboidal, or spindle-shaped, each having a large round nucleus and stained with eosin. These cells may be closely associated with chromaffin cells. In fishes treated with drugs and hormones or subjected to hypophysectomy (Surgical removal of the pituitary gland) or salinity changes, the inter-renal cells exhibit changes in shape and structure. These cells show vacuolisation (**Vacuolization** is the formation of vacuoles within or adjacent to cells) and reduction in size in a hypophysectomised fish. Temperature shock, treatment with heavy metals and maturation of gonads cause histological alterations in the inter-renal cells of the fish.

Two kinds of secretions are produced by the inter-renal cells. These are

1. Glucocorticoids that influence blood sugar
2. Mineral corticoids affecting osmoregulation

The inter-renals of teleosts are homologous with the adrenal cortex of mammals and secrete corticosteroids. However, their level in the blood varies considerably depending on the species, season, and physiological state of the fish. Corticosteroids are important for the regulation of water and electrolyte balance and the metabolism of carbohydrate and proteins. The inter-renals play a vital role in the osmotic regulation of the fish. They are involved in glucogenesis (The formation of glucose through the breakdown of glycogen). Experiments have shown that inter-renal ectomy in eels produces hypoglycaemia (hypoglycemia, also called low blood glucose or low blood sugar, which occurs when blood glucose drops below normal levels) and causes the loss of liver glycogen. It has also been shown that inter-renals are directly related to gonad development and maturation and are involved in stress and spawning migration.

2.3.5. Chromaffin tissue

Chromaffin cells (adrenaline-producing) are generally dispersed in the head kidney near the postcardinal vein and exhibit considerable variation in their distribution in teleosts. These cells are rounded in appearance with slightly basophilic cytoplasm and may show typical chromaffin reactions also. The chromaffin cells may occur singly or in groups in different locations in the head kidney. They may be

1. Embedded in the vein wall or
2. Associated with the inter-renal tissue or
3. Dispersed in the inter-renal tissue and the vein walls
4. Present in the ovary of teleosts.

Studies show that 2 hormones, adrenaline and non-adrenaline, are present in the head kidney and the plasma of teleosts but vary in their concentration in different species. Fish treated with adrenaline immediately become hyperglycaemic (Very high blood sugar). Chromaffin cells of teleosts contain dihydroxyphenylalanine (dopa) and 5-hydroxytryptamine (serotonin). Extracts from the chromaffin tissue stimulate the sympathetic nervous system of another fish. The hormones produced by the chromaffin cells, epinephrine and nor-epinephrine, perform functions comparable to the adrenal medulla of mammals. It concentrates pigment granules in melanophores and controls blood pressure. Hence, chromaffin tissue of teleosts may be considered homologous to the adrenal medulla of mammals. Chromaffin tissue is reported to be present in the ovary of several freshwater teleosts like *L. rohita, Catla, Mrigala, Puntius spp, Channa spp,* and *Notopterus* inside the ovarial folds along the lamellae.

2.3.6. The ultimo-branchial gland

The ultimo-branchial gland originates from the pharyngeal epithelium of the last or 'ultimate' gill pouch. It is unpaired. In chondrichthys, an ultimo-branchial gland is present between the pericardium and pharynx. In bony fishes, the gland is present between the abdominal cavity and sinus venosus ventral to the oesophagus.

In teleosts, the gland is seen as a band of white tissue on the septum (In anatomy, a **septum** is a wall dividing a cavity or structure into smaller ones) and may be solid or diffused. The gland may be paired or single and present a variable histological structure superficially resembling the parathyroid of higher vertebrates.

Parathyroid gland – It is an endocrine organ usually associated with the thyroid gland and possessed by all vertebrates except the fishes.

This ultimo-branchial gland secretes calcitonin (lowers serum calcium levels) that acts with hypocalcin (a protein secreted by the corpuscles of stannius called teleocalcin) to regulate calcium metabolism. Calcitonin is a hormone which lowers blood Ca^{2+} levels. In contrast, this hormone is secreted from the thyroid gland in mammals. Hypocalcemia (lowering of plasma calcium and phosphorus) is observed when calcitonin is administered to catfish and eel. Removal of the gland causes hypercalcemia. The gland may also be involved in the migration and reproductive behaviour of fish.

2.3.7. Islets of Langerhans or Endocrine Pancreas

Pancreas is a mixed gland. It comprises both the exocrinal and the endocrinal parts. The endocrine pancreas is present in most fish as islets of Langerhans or Brockman bodies. It arises from the endodermal lining of the embryonic foregut. Islet of Langerhans is present near the gall bladder or spleen. In some species, the islets are very large and may be grossly visible. During the spawning season, the size and number of islets in some fish will increase. It consists of 3 types of cells- alpha cells, beta cells and D cells. Alpha cells secrete glucagon hormone, which causes hyperglycemia. Insulin released by beta cells also controls glucose levels, and the function of D cells are still unknown.

2.3.8. Corpuscles of Stannius (CS)

Corpuscles of stannius (CS) is a unique endocrine organ in fish. They are found only in bony fishes. They were first observed by H. Stannius in 1839. The corpuscles of stannius are small nodular bodies present wholly or partly embedded on the dorsal or, dorsolateral or ventrolateral sides of the kidney in teleosts. Generally, these glands are present in the posterior portion of the kidney.

The number of stannius corpuscles varies in different fishes like *Channa, mystus, Catla,* etc. The number of corpuscles of stannius is usually 2 or more than 2, but in most fishes. This organ secretes a protein called hypocalcin (teleocalcin) that acts with calcitonin to regulate calcium metabolism.

The endocrine function of the corpuscles of stannius is still doubtful. However, the corpuscles of stannius perform the following functions.

Presence of renin activity: Extracts of corpuscles of stannius of the eel increase blood pressure, resembling that of renin. (Renin is a protein (enzyme) released into the bloodstream by special kidney cells. It is released due to decreasing salt (sodium) levels or low blood volume. Renin plays a role in releasing aldosterone, a hormone that helps control the body's salt and water balance).

Osmoregulatory function: Removal of the corpuscles of stannius from the eel caused a reduction in the plasma sodium and increased potassium and calcium concentration, which may even result in death. So, corpuscles of stannius affect osmoregulation.

Effect on kidney function: The kidney is the primary organ upon which the secretions of the corpuscles of stannius show maximum activity. Removal of corpuscles of stannius did not affect the sodium transporting system in the gills of either freshwater or seawater eels but marked changes in renal function as evident from a decline in the urinary calcium and magnesium excretion.

2.3.9. The Urophysis (Caudal neurosecretory System)

Most fish also possess a urophysis, a neural secretory gland similar to the posterior pituitary but located in the tail and associated with the spinal cord. Urophysis consists of neurosecretory cells.

The hormones of urophysis are called "urotensins", and four kinds of them are identified. These are urotensin I, II, III and IV. These are peptides; all 4 may not be present in the same fish. However, urotensin I and II are commonly found in fish, and the central nervous system controls their release. The exact function of urophysis is still not known. However, the extract of urophysis produced the following effects.

Urotensin I: It increases the blood pressure of fish but is less potent than urotensin II.

Urotensin II: It is involved in the contraction of smooth muscles such as the urinary bladder, causes a marked increase in blood pressure and enhances urine flow.

Urotensin III: It induces the sodium intake across the gills (osmoregulation) of goldfish. However, the effect of this component is not observed in other fishes.

Urotensin IV: It shows activity like antidiuretic hormones of the pituitary gland.

2.3.10. Gonads

The hypothalamus produces **gonadotropin-releasing hormone (GnRH)**, which is thought to stimulate the pituitary gland, a small gland located beneath the brain, to produce and release **gonadotropin hormones (GtHs).** Gonadotropin hormones (GTH) act on the ovaries and testes **(gonads).**

The hormones of the reproductive system of vertebrates (sex hormones) are steroids. In fish, sex hormones are produced by the gonads, which control the maturation and development of secondary sexual characteristics. In male fish, hormones are believed to be secreted by the testis's interstitial cells, which are involved in steroidogenesis. In the female, the post-ovulatory corpus luteum appears to be the source of sex hormones.

LH stimulates the testes to produce several kinds of steroid hormones called ***androgens***. One of these androgens is ***testosterone***, the primary sex hormone in males. LH stimulates the ovaries to produce ***estrogen*** and ***progesterone***, the female sex hormones.

Gastro-Intestinal hormones

Secretion of gastric and pancreatic juice is regulated by hormones secreted by the intestinal mucosa in mammals and other vertebrates. These hormones are gastrin, secretin and cholecystokinin (CCK). In addition to these, a few other peptides have been identified. However, the position of fish is not known. Salmon and some other teleosts exhibit gastrin-like activity in the stomach. A CCK-like hormone is also reported to be present in fish, but its function is unknown. A third hormone, secretin, stimulates the rate of pancreatic juice secretion.

2.3.11. Pineal gland

The pineal gland is a light-sensitive neuroendocrine structure in the anterior brain and is a well-vascularized organ. This gland secretes melatonin, which may play a role in controlling reproduction, growth, and migration.

2.4. Pheromones

Pheromones are a type of hormone that is mainly used to attract others. About 50 years ago, Karlson and Lu¨scher coined the term 'pheromone' to describe chemicals that are 'excreted to the outside by an individual and received by a second individual of the same species in which they release a specific reaction.' Pheromones are chemical substances secreted by animals in small quantities and cause various intraspecific responses. In fish, pheromones are released into the surrounding water, and other individuals of the species quickly pick up the information. Only minute quantities of pheromones are released into water, and this helps in:

1. Recognition of species and sex.
2. Recognition of territory
3. Warning and social behaviour

Pheromones play an essential role in the recognition of species and sex. Experiments have shown that males of some species, like *Ictalurus* and *Brachydanio renio*, are attracted by the odour of females. In some species, the odour of the ripe female attracts the male, while in others, the odour of the mature male fish attracts the female. These chemicals or attractants are produced by the gonads and are released into the water directly or through urine.

Recognition of territory is also an important function of pheromones. Many species of fish are capable of conditioning the surrounding water by releasing pheromones, and other members of the species recognise this. It has been suggested that pheromones influence the homing behaviour of migratory species. Thus, the adults of Atlantic salmon, *Salmo salar*, return to the same river and reach the same area by recognising the odour left by young fish.

Pheromones produced by special club cells of the skin's epidermis produce an alarm reaction. Thus, the club cells are broken and release the pheromone when a predator catches and eats a fish. This information is immediately picked up by the olfactory organs of other group individuals, who scatter and escape from the enemy. The chemical nature of these pheromones is unknown, and these may be amino sugars. Besides the alarm or fright reaction, the pheromones may influence other fish behaviours, such as schooling. All the species may not secrete the alarm substance, but it is apparent that the pheromones play an important role in guiding the behaviour of the fish.

2.5. Endocrine glands of Shellfishes - Crustaceans

The endocrine systems of crustaceans resemble those of insects.

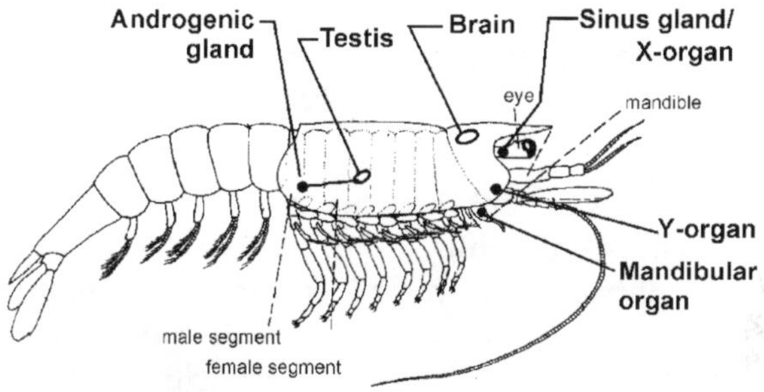

Fig 4.9 - Endocrine glands of Decapod Crustaceans

The major neuroendocrine control centre of malacostracans is the X-organ–sinus-gland complex. X- organ is a specialised neurosecretory organ in both the eye stalks. The optic peduncle has 3 ganglionic (A ganglion is a nerve cell cluster or a group of nerve cell bodies located in the autonomic nervous system and sensory system.) formations such as medulla externa, medulla interna, medulla terminalis. X-organ is located in the medulla terminalis. The axons (the long thread-like part of a nerve cell along which impulses are conducted from the cell body to other cells.) of these secretory neurons and neurohaemal organs are together called the sinus gland, which serves as a storage–release centre for these hormones. As the sinus gland serves for the storage and release of neurohormones secreted by secretary neurons of the x-organ, it is called X- organ-sinus-gland complex

The X-organ passes its secretions to the sinus gland, which acts as a release centre into the blood.

X-organ/SG system - production and release site for the various eyestalk hormones

- Ovary (gonad) Inhibiting Hormones (OIH/GIH)
- Moult Inhibiting Hormone (MIH)
- Light adapting hormone or distal retinal pigment hormone
- Hyper Glycemic Factor (CHH- Crustacean Hyperglycaemic Hormone)
- Erythrophore concentrating hormone

- Neuro depressing hormone.
- Growth Stimulating Hormone (GSH) is also known as Moult or Gonad Stimulating Hormone. The Thoracic ganglia secrete it.

This complex regulates maturation, dispersal of pigments in the eye and body colour change. They also regulate molting and the associated metabolic functions by actions exerted upon the maxillary Y -organ in the head. In crustaceans, neurosecretion inhibits secretions from the Y-organ, and the withdrawal of the inhibitory hormone initiates the molt.

2.5.1. Y organ

Y organ is an endocrine organ located in the head. It produces moulting crustecydsone, identical to b-ecdysone (insect moulting hormone –prothoracic glands).

Panouse (1943) observes that eyestalk ablation enhances ovarian growth, vitellogenesis, and oviposition. Thus, reproduction in crustaceans appears to be under the inhibitory control of an eyestalk factor known as the gonad inhibiting hormone (GIH), which may be the same as the "vitellogenin inhibiting hormone (VIH). The eyestalk is a source of many other hormones including the "moult inhibiting hormone (MIH) which directly inhibits ecdysterone production by the Y-organs in culture.

2.5.2. Mandibular organ

The mandibular organ in crustaceans is similar to that of an endocrine gland in higher vertebrates. It secretes hormones or hormone-like substances that influence the reproductive organs or reproductive process. It has been involved in ovarian growth and vitellogenesis. The mandibular organ synthesizes methyl farnesoate (MF), responsible for juvenile characteristics. MF has been shown to have a stimulatory effect on the Y-organ, which secretes ecdysone.

Decapods, for example, have a unique organ located in the walls of the pericardium enclosing the heart; the pericardial organ secretes a substance, perhaps a polypeptide neurohormone, that accelerates the heartbeat.

The female's ovaries, the male's reproductive glands, the pericardial organs, and the Y-organs of decapods also produce hormones that function in the molt and reproductive cycles.

Neurosecretory hormones of crustaceans have diverse chemical and biological characteristics but are polypeptides, as are the neurosecretory hormones of vertebrates.

The sinus gland comprises the terminals from neurons, which have cell bodies in the X-organ and brain.

Unlike insects, crustaceans have an androgenic gland, which typically is located on the genital duct (vas deferens) of the male. The androgenic gland secretes a hormone, possibly steroid in nature, which controls both the differentiation of the gonad of the male into a testis and the male characteristics of its limbs. The absence of the androgenic gland in the female results in an ovary formation that subsequently synthesizes one or more hormones. These promote the development of brood chambers (in which the young are hatched) and other structures associated with reproduction in female amphipods.

2.6. Molluscan Hormones

Hormone production is not well documented in molluscs other than gastropods and cephalopods. Antagonistic neurohormonal control of reproductive activity and metabolic processes is performed in the gastropods through cerebral dorsal bodies and lateral lobes or juxtaposed organs. A pair of salivary glands line the oesophagus, a large digestive gland called the liver empties into the stomach and a gland secreting albumin.

Experimental studies indicate an endocrine relationship in gastropods between the gonad (ovotestis) and possible neurosecretory cells in the tentacles and the brain; one ganglion of the gastropod *Lymnaea* may secrete a neurohormone with a diuretic (urine-producing) action.

Epithelial glands in molluscs are important; in the cephalopods, which are the most advanced invertebrates in some respects, optic glands on the optic stalks (eyestalks) secrete a hormone that promotes the development and maturation of the gonads. In immature cephalopods, the activity of the glands is inhibited by the central nervous system, apparently by a chemical mediator that diffuses from nerve fibres. The anterior salivary gland secretes the clear liquid (of water, mucin, protein and enzymes) into the mouth. It moistens food and starts the breakdown of starches. The posterior salivary gland (poison gland) produces venom to paralyze prey. The oviductal gland surrounding the end of the primary oviduct is responsible for secreting some of the external coatings over spawned eggs. In octopus, it also acts as a spermatheca. In some cephalopods, the hormones also affect death by starvation after the molluscs have deposited eggs or mated.

The nerve net, which constitutes the very primitive nervous system of the coelenterates, probably the most primitive multicellular animals, apparently contains neurosecretory cells; indirect but convincing evidence suggests that the cells release a secretion that promotes growth and inhibits sexual reproduction.

Neurosecretions by cells outside the nerve cell bodies (ganglia) have been described in gastropods and cephalopods, the released hormones diffusing through the tissues rather than being concentrated in special organs.

UNIT V

CHAPTER 1

EFFECT OF ENVIRONMENTAL FACTORS ON THE PHYSIOLOGY OF FISH

1.1. Major environmental factors that affect the physiology of fishes

- Temperature
- Salinity
- DO
- pH

1.1.1. Temperature

A slight increase in temperature is beneficial to physiological rates, resulting in more energy, higher rates of diffusion, and more enzyme-substrate complexes that produce higher reaction rates for growth, maturation, etc. However, higher temperatures beyond tolerance limits negatively impact physiology, as cardiac output cannot keep pace with increased metabolic demands.

Response to increase in temperature:

- Loss of plasma electrolyte (dehydration) and shift of water out of cells which increase the extracellular volume
- Change in spawning activity:
  - » Disrupt the gonadal maturation in both genders.
  - » Low levels of plasma sex steroids
  - » Reduced levels of male 11-ketotestosterone cause deleterious effects on spermatogenesis
- Increase heart rate and cardiac output

Response to decrease in temperature

- Loss of plasma chloride
- Decrease in heart rate
- Delay in sexual maturation
- Slow growth and feed conversion ratio

Response to change in salinity

- Change in salinity affects egg fertilization, incubation, yolk sac resorption, early embryogenesis, swim bladder inflation, and larval growth.
- Increase secretion of oxytocin and vasopressin
- Increase in osmoregulation activity
- Decrease in overall growth
- Decrease in feed conversion ratio
- Alteration in intestinal bicarbonate secretion
- Exposure to hyper-salinity increases characteristic traits related to osmoregulation exhibited by fish in seawater (e.g., increased drinking rate, plasma osmotic pressure, intestinal absorption and plasma concentrations of Na^+, Cl^- and K^+. Fish exposed to high salinity also increase HCO_3^- ATPase activity in the gut and gill).
- Environmental salinity also influenced the relative appearance of lactate and metabolic acid in the extracellular fluid compartment, with full-strength salinity favouring the relative appearance of lactate in the blood.

Fish response to hypoxia

- Hypoxia or oxygen depletion is a phenomenon that occurs in aquatic environments as dissolved oxygen (DO; molecular oxygen dissolved in water) becomes reduced in concentration to a point detrimental to aquatic organisms living in the system.
- At low DO, the blood flow can be increased by opening up further secondary lamellae to increase the effective respiratory area, and the concentration of red blood corpuscles can be increased to raise the oxygen-carrying capacity of the blood per unit volume. The latter can be achieved by reducing the blood plasma volume (e.g., by increasing the urine flow rate) in the short term and releasing extra blood corpuscles from the spleen in the longer term.
- Ventilation rate is increased ventilation rate is increased to bring more water into contact with the gills within a unit of time
- The major pathological-anatomic changes include a very pale skin colour, adherence of the gill lamellae, and small haemorrhages in the front of the ocular cavity and the skin of the gill covers.

- Fish reduce food intake, leading to a reduction in growth.
- Reproduction is inhibited, and fertilization success and larval survival are compromised.
- Energy utilization is decreased, associated with a shift from aerobic to anaerobic metabolism.
- Transcription is reduced, mediated by increased levels of hypoxia-inducing factor 1 (HIF-1), which also up-regulates genes involved in erythropoiesis, capillary growth and glucose transport.

Fish response to hyperoxia

- Hyperoxia is the state of water when it holds a very high amount of oxygen.
- If fish are exposed (at a lower atmospheric pressure) to such water, their blood equilibrates with the excess pressure in the water. Bubbles form in the blood, blocking the capillaries; in sub-acute cases, the dorsal and caudal fins can be affected, and bubbles may be visible between the fin rays.
- In severe cases, death occurs rapidly as a result of blockage of the major arteries
- The epidermal tissue distal to the occlusions then becomes necrotic.
- When the water is supersaturated (hyperoxia), the bladder becomes over-inflated, and this leads to buoyancy problems, especially in small fishes

Response to low pH

- Levels of pH 4 constitute acidic conditions lethal to most fish species.
- When fish were exposed to low pH, loss of sodium (Na^+) and chloride (Cl^-) ions from the body fluid occurred, resulting in a decrease in plasma osmotic pressure.
- At low pH, gill chloride cells showed high V-ATPase activity that stimulates the proton pump of the cell membrane to excrete H^+ ions.
- Decrease in pH also reduces the oxygen-carrying capacity of haemoglobin.

  Even if pH is not low enough to be lethal, the stress of acidification induces various physiological and ecological problems in fish.
- At low pH, plasma cortisol levels peaked in response to acid stress, and immunoglobulin (IgM) levels subsequently decreased. This suggests that acid stress depresses the immune system of fish.
- At low pH, plasma levels of sex steroids and gonadotropin exhibited abnormally high levels, and there was a possibility that acid stress disrupted the endocrine control over reproduction.

CHAPTER 2

STRESS PHYSIOLOGY

2.1. Stress – Definition

Stress can be described as the physiological response to a stressor. In other words, stress is an internal physiological state that is caused by external conditions. Stress can also be described as an internal hormonal response of a living organism caused by environmental or other external factors that move that organism out of its normal physiological resting state or homeostasis. Stress can disturb the normal physiological equilibrium or homeostasis of fish by forcing the reallocation of energy within its system.

Stressors are those factors that cause a stress reaction. They may be crowding, low dissolved oxygen levels, high carbon dioxide levels, photo-stress, thermal stress, pathogens, heavy metals, toxins, handling, aggression by other fish, electrical charges, excessive water velocity, nitrogenous waste, low pH, noise or vibration, nutritional deficiencies, transport stress, improper hardness, injury, etc.

2.2. Osmotic stress

Particular emphasis is placed on the physiological effects of the osmotic stress of transferring fish from freshwater to marine environments, and some stress occurs in reverse. During this time, some stress occurred, called Osmotic Stress.

2.3. Thermal stress

Thermal stress is defined as any temperature change that produces a significant disturbance in the normal function of a freshwater teleost, decreasing the probability of survival. Thermal stress is considered only at the level of individual fish, and emphasis is placed on ecological rather than physiological effects.

2.4. General Adaption Syndrome – Stages of Stress Response

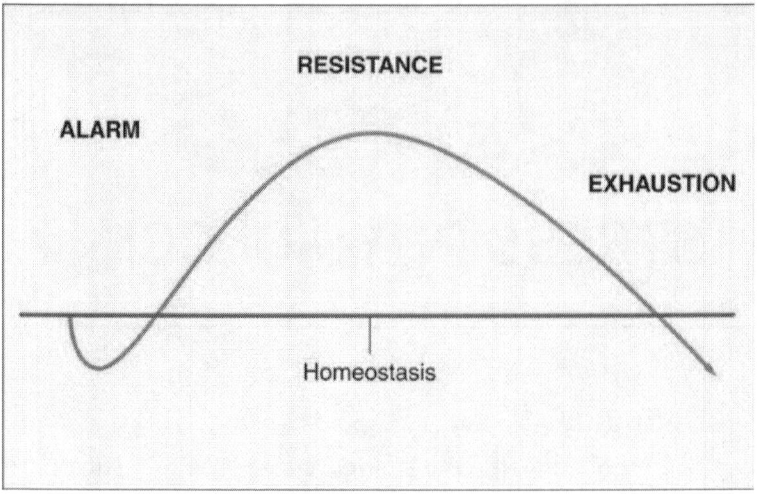

Fig 5.1 - Stages of Stress Response

Stage of alarm

Sympathetic NS of ANS is activated, and the adrenal gland secretes the stress hormone cortisol, along with adrenaline

Stage of resistance

Parasympathetic NS of ANS is activated, the results of the hormonal changes, which occurred in the previous stage are still take

Stage of exhaustion

If the stressor is removed, the body will start to recover. Otherwise, due to the depletion of resources, animals may die or adapt.

2.5. Types of Stress Response

Fish responses to stress can be divided into three phases: primary, secondary, and tertiary.

The **primary phase** refers to a generalized neuroendocrine response in which catecholamines (epinephrine and norepinephrine) and cortisol are released from chromaffin and interrenal cells.

In **secondary phase,** higher circulating levels of these hormones trigger a secondary response that involves physiologic and metabolic pathways; examples of the secondary response include hyperglycemia due to enhanced glycogenolysis and gluconeogenesis, vasodilation of arteries in gill filaments, increased cardiac stroke volume, and immune function depression. The first two phases are considered adaptive and enable fish to adjust to stressors and maintain homeostasis.

Tertiary responses involve systemic changes in which animals may become incapable of adapting to stressors, leading to adverse effects on their overall health, including their performance, growth, reproduction, disease resistance, and behaviour.

2.6. Internal hormonal response to stress

The internal hormonal response to stress in fish is similar to that of mammals. The most widely accepted indicator of stress is the blood plasma cortisol level. The two significant actions of cortisol in fish are regulation of hydromineral or osmotic balance and energy metabolism. Some consider the role of corticosteroids to be protecting the body from overdoing its defence mechanisms.

Corticosteroid and catecholamine hormones are released in response to stressors to adapt to or avoid the cause of stress. Although both stress hormones are released into the bloodstream as part of the integrated stress response to acute and chronic stressors, they play different roles. Catecholamines, or CA, are the primary hormones released in the "Fight or flight" response to acute stressors. Corticosteroids, predominately cortisol, are the primary hormone released in response to chronic stressors.

The catecholamine hormones epinephrine (adrenalin) and norepinephrine (noradrenaline) are associated with more immediate reactions to stress and are released when situations requiring a fight or flight response occur. The release of catecholamine hormones into the bloodstream causes increases in cardiac output, blood sugar, respiration, oxygen uptake, and blood flow to the gills. This prepares fish to better cope with threats to territory and safety. These responses are usually short and are considered acute stresses.

Corticosteroids, primarily cortisol, are associated with chronic stress as the animal attempts to adapt. Cortisol is released in response to all stressors, but as the stress reaction continues, its effects become more significant.

2.7. Effect of Catecholamines on the physiology of fish

Increase in blood glucose, increase in epithelium permeability to water and ions, increase in heart rate, increase in calcium reabsorption in bones, increase in cardiac output, increase in HCl secretion in the stomach, Increase in respiratory ventilation, increase in insulin release by pancreas, vasodilation and vasoconstriction of arterioles, increase in thyroxin release, increase in number of thrombocytes, decrease in protein synthesis in liver, increase in gluconeogenesis in liver, decrease in glycogen synthesis, increase in production of steroid hormones and smooth muscle tension.

2.8. Effect of Corticosteroids on the physiology of fish

Protein Metabolism - Increased protein synthesis via enhanced activity of messenger RNA. The protein synthesized seems to be primarily proteolytic. Inhibition of growth and mobilization of protein from thymus, spleen and liver

Carbohydrate Metabolism - Reduced utilization of carbohydrates. Increased glucose production from tissue protein. Deposition of glycogen in the liver. Mobilization of leucocytes. Reduced inflammatory response

Gills (in freshwater) - Increased retention of ions

Kidney (in freshwater) - Increased reabsorption of Na^+ by kidney tubules

Testes (in freshwater) - Precocious sexual maturity in downstream migrant smolts (males only)

Gills and intestine (in seawater) - Increased production and activity of Na^+/K^+ ATPase, enhanced salt absorption in the gut and enhanced salt excretion by the gills

2.9. Physiological effect of hypoxia

Hypoxia is a noticeable and dramatic effect of low ambient oxygen on an organism, a lethal response. Hypoxia may limit an organism's energy budget or scope for growth and activity. It may cause an organism to alter its behaviour and limit its tolerance to other environmental challenges.

Anaerobic Metabolism: In severely hypoxic water (low dissolved oxygen), most energy production is via anaerobic pathways, raising blood pH through increased lactate production.

Hypercapnic hypoxia: Hypercapnia, independent of dissolved oxygen, can dramatically affect the physiology of marine organisms. An elevation of ambient CO_2 results in a concomitant elevation of the CO_2 levels in the bodies of organisms. The direct result of hypercapnia is a decrease in the pH of tissues and body fluids, which can profoundly affect several functions.

Production of Reactive oxygen species (ROS) also has a negative effect on an organism's health.

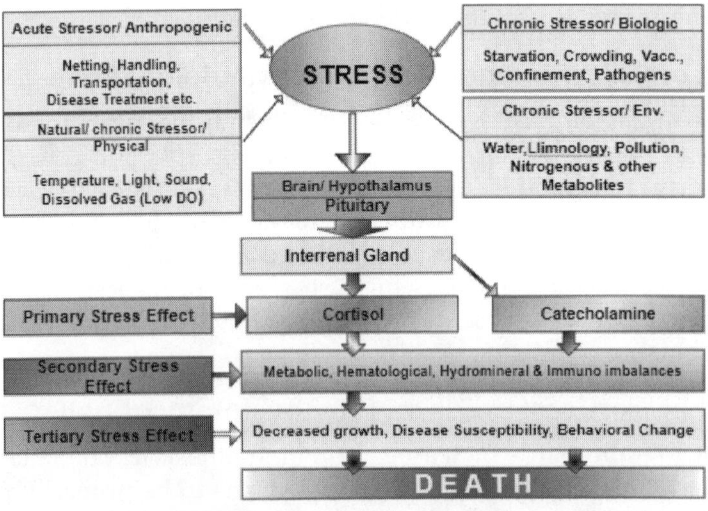

Fig 5.2 – Stress Model in Fish

REFERENCES

Ahmed, I., Jan, K., Fatma, S. and Dawood, M.A., 2022. Muscle proximate composition of various food fish species and their nutritional significance: A review. *Journal of Animal Physiology and Animal Nutrition*, *106*(3), pp.690-719

Fiedler, S., Wünnemann, H., Hofmann, I., Theobalt, N., Feuchtinger, A., Walch, A., Schwaiger, J., Wanke, R. and Blutke, A., 2020. A practical guide to unbiased quantitative morphological analyses of the gills of rainbow trout (Oncorhynchus mykiss) in ecotoxicological studies. *Plos one*, *15*(12), p.e0243462

http://www.rdscollege.ac.in/studymaterial/1596509204.pdf?uid=

https://afrilcate.com/percolation/

https://basu.org.in/wp-content/uploads/2020/03/Environmental-factors.pdf

https://basu.org.in/wp-content/uploads/2020/03/muscle-physiology.pdf

https://basu.org.in/wp-content/uploads/2020/03/shell-fish-sense-organ.pdf

https://biologyeducare.com/excretion-in-fish/#google_vignette

https://ecourses.icar.gov.in/

https://elearning.raghunathpurcollege.ac.in/files/1E5EFB8A16699054940.pdf

https://en.bdfish.org/2012/11/respiration-in-fishes/

https://en.wikipedia.org/wiki/Cardiac_pacemaker

https://i.ytimg.com/vi/0QVE1dFXFRo/maxresdefault.jpg

https://ib.bioninja.com.au/higher-level/topic-11-animal-physiology/112-movement/types-of-muscles.html

https://jinggasexchangejing.weebly.com/fish.html

https://jinggasexchangejing.weebly.com/fish.html

https://open.oregonstate.education/aandp/chapter/10-2-skeletal-muscle/

https://pressbooks-dev.oer.hawaii.edu/anatomyandphysiology/chapter/smooth-muscle/

https://slideplayer.com/slide/14184638/

https://www.alamy.com/urea-cycle-illustration-image529052450.html?imageid=2274ED0F-BF85-4ECE-8E7D-E74D24EDB14C&p=2085580&pn=1&searchId=be673eac6e94c6a6c649240fb2b1648f&searchtype=0

https://www.basu.org.in/wp-content/uploads/2020/03/Biochemistry-Water.pdf

https://www.basu.org.in/wp-content/uploads/2020/03/Stress-physiology.pdf

https://www.biologydiscussion.com/invertebrate-zoology/phylum-arthropoda/study-notes-on-true-crabs/33418

https://www.britannica.com/animal/crustacean/Form-and-function-of-internal-features

https://www.britannica.com/topic/fish-processing/Fat

https://www.differencebetween.com/difference-between-myofibril-and-vs-muscle-fiber/

https://www.fao.org/3/V7180E/V7180E04.htm#:~:text=A%20major%20difference%20is%2C%20further,O%20as%20the%20end%20products.

https://www.ncbi.nlm.nih.gov/books/NBK572070/

https://www.necropsymanual.net/en/teleosts-anatomy/digestive-system/

https://www.necropsymanual.net/en/teleosts-anatomy/skeleton-and-musculature/#:~:text=Appendicular%20muscle%20derives%20from%20trunk,to%20eye%20and%20gill%20movement.

https://www.notesonzoology.com/prawn/respiration-in-prawn-crustacean-anthrapods/1975#

https://www.parl.ns.ca/lobster/overview.htm

https://www.sciencedirect.com/topics/earth-and-planetary-sciences/hydrological-cycle

https://www.scribd.com/presentation/545764637/Crabs-Respiratory-System

https://www.yorku.ca/spk/fishbiol09/FB09lecture11.pdf

https://www.zoologytalks.com/mechanism-of-respiration-in-higher-invertebrates/

Walsh, T.S. and Saleh, E.E.D., 2006. Anaemia during critical illness. *BJA: British Journal of Anaesthesia*, *97*(3), pp.278-291

INDEX

D

E

F

G